TO THE LATE DENIS KEARNEY

WITH GREAT AFFECTION AND RESPECT

FOREWORD

These Proceedings consist of the contributed papers and conclusions of the Expert Panel on Low Consumption/Low Emission Automobile held in Rome on 14th-15th February 1990. The Expert Panel was conceived and organised as a part of the joint effort by the Organisation for Economic Co-operation and Development (OECD) and the International Energy Agency (IEA) in response to the priority accorded to transport, the environment and energy security by OECD and IEA Ministers. The Expert Panel considered technologies and policy options which can lead to the development and deployment on a large scale of highly fuel-efficient and less-polluting automobiles. In particular, the aims were to analyse and assess the various prospects for developing low consumption and low emissions automobiles, to suggest approaches in order to overcome technical, economic and other barriers to the market penetration of such automobiles, and to propose subjects for collaborative R&D and demonstration programmes in automobile technology.

The share of road transport in the oil consumption of OECD Member countries has grown rapidly in recent years, and in 1989 accounted for nearly 60% of OECD oil use. Automobiles account for the largest part of this consumption, and the expansion in car ownership is likely to accelerate in pace with economic growth, particularly in the non-OECD region. From an energy policy viewpoint, this rising consumption can ultimately have grave implications for the security of oil supply, which continues to be the most vulnerable element of the energy market. The environmental impact of atmospheric pollutants arising from automobiles, together with their contribution to congestion in cities, is a matter of great concern worldwide. Moreover, the ever-increasing emissions of greenhouse gases present the greatest challenge in formulating response strategies to global climate change. These problems require solutions which can accommodate the continuing contribution of the automobile industry to national economies and to international trade, and at the same time respond to the increasing popular demand for mobility. Recent advances in the automobile industry and elsewhere indicate that technology options which could substantially improve automobile fuel efficiency and emission levels are either available or already at the planning or development stage. The size and timing of these potential improvements are critically dependant for their realisation on appropriate governmental actions.

These Proceedings contain a wide variety of ideas and suggestions for immediate and longer-term actions. Consensus views were not sought on all issues considered by the Expert Panel, containing as it did a broad range of energy, environmental and transport expertise. However, the technical papers and the discussions which are summarised in these Proceedings represent a significant contribution on this important subject. While these do not necessarily reflect the views of the OECD, the IEA, or their Member governments, it is hoped that these Proceedings will be useful in promoting further analysis and ultimately actions leading to rapid reduction in the fuel consumption of automobiles while responding to continuing and emerging environmental concerns.

Bill L. Long
Director
Directorate for Environment
OECD

Sergio F. Garribba
Director
Energy Technology, Research
and Development
IEA

TABLE OF CONTENTS

APPENDICES

PART I

HIGHLIGHTS OF THE DISCUSSIONS OF THE EXPERT PANEL

1. BACKGROUND

The relationship between technology, energy and the environment is now more than ever a challenging issue for the OECD and the IEA Member countries, and presents a number of crucial elements in reconciling goals of energy security, environment protection and sustainable growth. The interplay between these elements is assuming particular importance in the transport sector. New and unanticipated influences are emerging which may fundamentally alter and re-shape this sector, in particular concerning the development and use of the automobile.

A steadily increasing portion of oil consumption and atmospheric emissions in the OECD region and other countries arises from road transport. In OECD Member countries, for instance, the share of the transport sector in total oil consumption has been rising from 38 per cent in 1973 to nearly 60 per cent in 1989, with road transport now accounting for about 85 per cent of total transport fuel use. Automobiles and light vehicles have by far the largest share of this rapidly increasing consumption, and this trend poses a serious threat for future energy and environmental policies. Motor vehicles are today responsible for a larger amount of atmospheric pollutant emissions than any other single human activity. Conventional pollutants such as nitrogen oxides, carbon monoxide, volatile hydrocarbons, particulates, toxic emissions including lead and benzene are a matter of concern, together with growing traffic congestion in many cities worldwide. Moreover, in the context of global climate change, the ever-increasing emissions of greenhouse gases - carbon dioxide (CO_2) in particular - pose a very critical problem. From an energy security viewpoint, rising fuel consumption by transport directly affects oil supply, which continues to be the most vulnerable element of the energy market.

During their meeting of 30th May 1989, the Energy Ministers of IEA Member countries called sharp attention to two aspects of the current energy situation to which they attached particular importance and viewed with deep concern: "growing world-wide oil consumption, particularly for transportation, with its eventual medium-term consequences of a tighter supply/demand balance and renewed vulnerability to supply disruptions; and the environmental aspects of energy supply and consumption, including both the more conventional and better-known pollutants and the growing atmospheric concentration of CO_2 and other "greenhouse" gases and its long-term consequences for global warming and climate change." Ministers also "agreed that greater energy efficiency and conservation is both possible and necessary in the use of all forms of energy. It will be pursued vigorously by the governments of all Member countries for both energy security and environmental reasons, concentrating on achieving long-term results in large and fast-growing sectors of energy use such as transportation ...."

These views were reiterated by the OECD Ministerial Council of 30th-31st May 1990 which also indicated its concern about possible global climate change, as well as the importance of OECD/IEA collaboration in developing policy options in the field. The transportation and electricity sectors were judged to be of particular concern, as they constitute the fastest growing end-use sectors and the major sources of emissions, especially those related to the accumulation of greenhouse gases in the atmosphere.

On 9th-11th July 1990, the Heads of State and Government of the seven major industrial democracies and the President of the Commission of the European Communities, at their annual meeting which took place in Houston, re-stated their commitment to undertake common efforts to limit emissions of greenhouse gases. They acknowledged the priority which must be given to improvements in energy efficiency, the importance of involving the private sector in the development of solutions to environmental problems, and encouraged the OECD to accelerate its work on environment and the economy.

The above statements are a clear indication to all entities concerned with automobile design, production and use that substantial reductions are necessary in the fuel consumption and atmospheric emissions of future automobiles. It is evident that governments and automobile manufacturers have, from time to time, addressed these objectives with varying degrees of success. Indeed, in some OECD Member countries the achievement of low fuel consumption has been a priority and a significant element in the competition between automobile manufacturers. It is clear, however, that an increased urgency is now required in the development of a new generation of automobiles having the lowest possible levels of fuel consumption and atmospheric pollution.

The development of highly fuel-efficient and less-polluting vehicles has been proposed as a means of achieving considerable benefits from environmental, socio-economic and energy security viewpoints, both for OECD countries and also for the newly industrialising countries, in view of their demand for increased mobility and rapidly growing automobile markets. A wide range of opinions and projections on this subject have been expressed by policy makers, individuals and entities concerned with transport, energy, technology, and environment, often in isolation from each other and without an opportunity to debate and resolve conflicting views. To provide an opportunity to examine these and other related issues, an Expert Panel meeting was jointly organised by the OECD Environment Committee (and its Group on Energy and Environment) and the IEA Committee on Energy Research and Development (and its Working Party on Energy End-Use Technologies) with contributions from the IEA's Standing Group on Long-Term Co-operation. The European Conference of Ministers of Transport and the Commission of the European Communities also took an active part in this initiative. The hosting organisation was the Italian Agency for Nuclear Energy and Alternative Energy Resources (ENEA).

The Panel consisted of experts from national and international organisations, institutes and R&D centres, and included specialists from major OECD car-manufacturing countries, so as to achieve a desirable balance of interests and interdisciplinary experience on the various questions of energy policy and technology, transportation, and the environment.

2. ISSUES CONSIDERED BY THE EXPERT PANEL

The Expert Panel considered issues affecting the entire range of automobiles: small, medium and large models, as well as light commercial vehicles. In relation to vehicle emission control and increased fuel efficiency, and relevant national and international experiences, the experts were invited to discuss:

(i) key factors and trends in automobile conception, design, manufacture and use which influence and determine fuel consumption and atmospheric emissions;

(ii) various near-term options arising from available technology comprehending: trends in automobile development and desirable improvements; present status of technology; energy conservation and security aspects; environmental goals; role of inter-industry competition, trade and other constraints;

(iii) approaches and options to promote, on a large-scale, more energy efficient and less polluting automobiles and overcome technical, economic and other barriers which might curb penetration of new or improved technology in OECD countries and worldwide;

(iv) mid- and long-term technology progress towards increased energy efficiency and reduced emissions; possible subjects for collaborative R&D and demonstration programmes in automobile technology;

(v) strategies for the development and deployment of improved automobiles; areas of opportunity for co-ordinated actions involving OECD Member countries and other entities.

Experts were also invited to debate how to improve environmental performance of vehicles in use. Approaches which were discussed included better inspection and maintenance programmes, use of diagnostic and retrofit mechanisms, and observance of speed limits. Further issues for consideration referred to fuels, and to efforts which could be made to improve the quality of fuels and to prepare substitutes. Finally, questions concerning the evolution of environmental regulations, their possible harmonization and impact on industrial competitiveness, the use of economic and pricing instruments, and the pace of technological progress were addressed by the experts. The problems of developing and promoting public transport, traffic management and urban policies were frequently raised during the meeting, as they are closely related to the questions debated. However, the attention of the Expert Panel focused on the reduction of automobile fuel consumption and emissions from a technological, regulatory and economic viewpoint.

It should be noted that because of the diversity of the experts participating, a wide range of often conflicting views was evident. On the most significant issues where the experts could not achieve a relative consensus view, the aim was to record the range of opinions expressed. Possible future actions were identified which could be addressed in more depth at subsequent meetings.

3. TECHNOLOGY OPTIONS TO REDUCE FUEL CONSUMPTION AND EMISSIONS

The experts discussed a wide range of technology options and improvements which are described in detail in the contributed papers included in the Proceedings. The discussions concentrated on the optimum combination of technology options in automobiles, and on the resulting improvements in fuel efficiency and emissions. Experts noted that the following trends appeared important in considering progress in automobile efficiency and emissions.

a) Specific fuel consumption of automobiles tends to remain high in contrast to the progress achieved in many other technological sectors. Since the early 1980s, the rate of improvement in fuel efficiency of the vehicle fleet has declined significantly, due principally to low oil prices and the increasing consumer demand for larger and more powerful cars. In addition, some types of emission control equipment have themselves lead to increased fuel consumption. This declining trend in fuel efficiency gains has occurred in spite of the availability of increased fuel-saving technologies which have emerged during this period.

b) Yearly mileage per vehicle is rising in several countries. This is a consequence of increasing personal mobility resulting from rising incomes and the low price of fuel. As a result of changing living patterns and the inadequacy of many mass transit systems, the portion of driving which takes place in congested urban and suburban areas is also increasing, leading to much higher fuel consumption per kilometre driven.

c) Although maximum speeds are now limited in most countries of the world, new models are often able to attain much higher speeds and such increasingly powerful cars consume more fuel, even at low speed.

d) The stock of vehicles is expanding worldwide. This arises mainly from socio-economic development and demographic growth. The total world vehicle fleet, presently of the order of 500 million, is expected to multiply by three or four within the next two decades or so, mostly due to increasing demand in the newly industrialising countries. As a result, this growth in vehicle numbers will quickly exacerbate the adverse consequences of delaying the introduction of improved automobiles.

e) The extended durability of automobiles due to better materials, construction and repair techniques may tend to prolong the time required for the deployment of significant numbers of improved vehicles in some markets.

The experts were invited to assess achievable progress in relation to the time needed for the deployment of vehicles which not only have very low fuel consumption but which also achieve desirable levels for pollutants such as nitrogen oxide, volatile hydrocarbons, carbon monoxide, particulates, and which produce less CO_2 emissions. It emerged from the discussions that significant improvements in fuel consumption and emissions can be achieved, even for vehicles which will come on the market in the relatively near-term, based on presently-available technology. The principal gains in fuel consumption can be obtained by focusing first on: more fuel-efficient engines; better transmission systems; vehicle weight reduction; aerodynamic improvement. Many car manufacturers have already designed and constructed advanced prototype automobiles that can offer considerable improvements in fuel consumption, but which, for commercial and other reasons have not yet been further developed and introduced in the market. Innovative systems for automobile propulsion and control have been proposed or are already at the development stage, that could lead to even greater improvements in energy efficiency and emissions reduction. These and other advanced solutions may be especially relevant for particular geographical regions, or for specialised patterns of vehicle use in, for example, congested urban areas.

The discussion by the experts indicated that the impact of potential improvements in fuel efficiency is difficult to quantify, since it depends on a large number of factors including: time horizon; types of vehicles; number, level and interaction of improvements; consumer acceptance; inter-industry competition and trade conditions; and above all, the level of incentives and pressure which governments will apply in this direction. In the short-term (1995-2000), significant gains in the fuel efficiency of new models (above that now available in existing fleets) were considered by many experts as achievable with today's techniques and fuels, while satisfying environmental regulations. These gains were thought to depend largely on the economic and regulatory incentives which governments may apply in order to ensure their uptake. A number of experts suggested that if these incentives are modest the resulting gains may also be modest (15-20 per cent improvement); if they are strong enough, the gains may be important (up to 40 per cent improvement). Such results could be obtained while to a large extent retaining the expected levels of comfort and interior space, together with satisfactory performance of the automobile (account being taken of speed limits existing in OECD Member countries).

Concerning car dimensions, aesthetics and comfort, there was a consensus that fuel economy did not necessarily imply "small" cars, but that the car park would continue to be composed of small, medium, and relatively large vehicles. Several experts judged that a shift to more reasonable dimensions was desirable for certain countries. Experts thought that the interior usable volume could be maintained in general, while the exterior volume and total weight could often be substantially reduced by appropriate design techniques. They further suggested that while customers would continue to demand improvements in comfort, drivability, luxury and aesthetics, these could be met while at the same time achieving reductions in weight, power, fuel consumption and emissions, provided effective fuel economy policies were implemented. Some experts remarked that different innovations which individually may reduce fuel consumption may not produce a proportionate cumulative effect when applied together. Nevertheless, cases may occur of improvements which provide multiplier effects or cascade gains in the overall car system. As an example, reduction in the weight of the car (optimised structure, lighter materials, etc.) permits a reduction in the power of the engine and thus its weight and consequently the weight of the transmission, and other parts of the drive train.

During the technical discussions, considerable time was devoted to the engine, as this is the crucial element for fuel economy and emission limitations. The experts discussed the possible options which are available to increase engine efficiency and flexibility of use, and to decrease its fuel consumption, weight and maximum power, while still maintaining sufficient performance when needed. Combining, for example, the use of high power density engines, turbo- or super-charging, electronic control of fuel injection and engine regulation, electronically controlled continuously variable transmissions, and so forth can permit the power of the engine to be maximised where needed, while retaining the fuel economy characteristics of a lighter and less powerful engine. Some of these options can be implemented now, while others may require additional development.

Some experts pointed out that in many countries, the potential efficiency gains which could have been accomplished in automobiles over the last ten years, have not been realised due to a shift towards more and more powerful engines. There were experts who noted that marketing efforts by many manufacturers were now directed towards emphasising the power, speed and acceleration of cars, in contrast to the emphasis on fuel economy which was current in the 1970s. This situation, it was said, would be detrimental to achieving improved levels of fuel economy and emissions, and that efforts were required from national and international authorities in order to counter this trend.

It was noted by the experts that the quest for large improvements in automobile fuel efficiency is not a new issue. A number of presentations referred to those automobile companies which have developed over the last ten years a number of full-size experimental vehicles (i.e. for 4/5 passengers), which achieved remarkable results (for instance, 2-3 litres/100 kms or 80-100 miles per gallon when measured over a representative driving cycle, while retaining the capability for maximum speeds over 150 km/hr or 95 miles/hr). While it is clear that the technical solutions selected for such experimental vehicles may not always be fully applicable on production models, experts felt that they can provide helpful experience and are useful in defining the limits of achievable progress. Nevertheless, due to the low fuel prices of recent years and lack of appropriate incentives, these efforts have not yet given rise to the development and large-scale production of highly efficient vehicles based on these prototypes

About 75 per cent of passenger journeys in many countries now occur on relatively short trips (less than 5km), often at reduced speeds in congested urban and suburban areas. The combination of cold starts (which greatly increases the fuel consumption for the initial few kilometres) and congested traffic (which may also double it or more) inevitably results in considerable increases in fuel consumption above that obtained in open highway conditions. In this context, the experts discussed technological solutions for reducing over-consumption, such as: increased use of electronics to augment the efficiency of the engine and overall drive train, optimisation of engine heating/cooling systems, engine stop and re-start systems for intermittent traffic flow, modular engines (disconnecting some of the cylinders at low power settings), flywheels for energy recuperation, energy regenerative braking, and so forth. Solutions based on the more promising of these technologies could be implemented relatively soon with proper incentives.

Progress which could be achieved in the medium-term using currently available technologies and fuels or technologies under development was reviewed and showed very promising prospects. In addition, the experts discussed the possible longer-term developments based on new concepts and technologies. Beyond the year 2000, the experts considered that the implementation of the best available technologies along with increased consumer demand for highly efficient and clean automobiles could lead to further gains in the fuel efficiency of cars, of the order of 50-60 per cent above today's levels.

Engine developments were discussed such as: compact and efficient "two stroke" engines with electronic fuel injection; gas turbine propulsion; efficient and clean "lean burn" engines for which problems of catalyst durability in conditions of poor maintenance could be overcome. The extent of the fuel consumption penalty of existing three-way catalysts (presently in the range of 5-10 per cent) was extensively debated, as well as the outlook for the development of catalysts for both diesel and "lean burn" engines. It was considered by some participants that breakthroughs could be expected relatively soon in the design and operation of these components. Experts noted the important contribution which new materials capable of withstanding extremely high temperatures can make to augment engine efficiency. The use of ceramic materials for components in contact with hot exhaust gases which is already taking place in some models was considered to have the potential for much wider application in the future. But some experts thought that this trend would not lead to the deployment of uncooled engines using monolithic ceramic components, due to problems of materials durability, cost, and the complexity involved in utilising the resulting higher energy levels in exhaust gas streams.

The prospects for existing and improved petroleum-based fuels was also considered. It would appear that up to the turn of the century, gasoline and diesel fuels will continue to meet the great majority of demand, as they are the only transport fuels likely to be available in sufficient volume and at acceptable prices. A number of experts claimed that important improvements in present gasoline and diesel fuel formulation could be quickly implemented and may result in increased engine efficiency while decreasing emissions from vehicles.

The experts considered the status and future viability of alternative fuels such as natural gas, methanol, ethanol, hydrogen and electricity. At present, the direct use of natural gas on a large-scale still poses problems of distribution and consequential limitation on vehicle range. In addition, the release of methane into the atmosphere which could result from multiple sources of leakages during transport, distribution and use might itself induce increasing greenhouse gas accumulation in the atmosphere. Yet natural gas was thought by some experts to have a promising future as a transport fuel, if the prevention of leakage can be assured. Experts noted that methanol poses complex problems as it diminishes some pollutants and increases others. Its use would be encouraged for some very polluted urban areas in the USA and also in Scandinavia where the use of methanol derived from biomass feedstock is thought to present a viable alternative to petroleum in certain circumstances. The methanol option was viewed by experts as rather limited in the near-term, due to the restricted availability and high cost of natural gas or biomass feedstocks as well as environmental difficulties with conversion processes using coal as feedstock. But in the future, conversion of methanol in fuel cells holds the promise of high efficiency along with low emissions. The use of ethanol produced from agricultural raw materials is subject to similar constraints as methanol and is not expected to develop in the near future while oil prices remain low. Experts noted that the use of hydrogen fuel introduces a number of technological problems which require much additional effort, such as: storage (the weight penalty of the tanks which is up to 15 - 20 times that of the stored hydrogen); production currently dependent on the availability of low-cost electricity; possibility of leakage and safety concerns. Long-term solutions to these constraints, in particular the development of low-cost production systems, perhaps based on advanced photochemical processes could make hydrogen a viable vehicle fuel for the distant future.

Progress in the use of electricity for vehicle propulsion has been slow due to the lack of a light, compact, durable, low-cost battery. However, even in the absence of breakthroughs for the battery itself, experts drew attention to promising solutions which are being developed, e.g. the "hybrid" car which combines an electric motor (more efficient and less polluting in urban conditions) and a conventional engine for highway traffic (which permits recharging of batteries). Some experts felt that such cars may soon appear on the market. The prospects for fuel cell propulsion in the medium to long term were thought to be promising, in light of the rapid advances in fuel cell materials and catalysts which are now taking place. Experts noted that the comparison of the environmental benefits of electric vehicles against other options would necessitate a full fuel cycle analysis of the increased electricity generation required. A trend towards market differentiation was forecast by some experts, in which specialised automobile types would be used as appropriate: for example, small electric vehicles for urban and suburban use, and larger conventionally-fuelled (and efficient) vehicles for longer distances. Innovative forms of car ownership, renting and sharing were suggested in place of multi-vehicle ownership in such circumstances. Electric vehicles in general were thought to have very promising medium- and long-term potential, in the context of this differentiated market, especially in urban areas, where policies may in the future progressively exclude other vehicles.

4. PROPOSED MEASURES FOR IMPLEMENTING DESIRABLE OPTIONS

The considerable environmental, energy security and socio-economic benefits arising from the development and market penetration of low-consumption/low-emission vehicles were discussed. These benefits could accrue not only to OECD countries but also to newly industrialising countries, in view of their rapidly growing automobile market. Policy tools such as new standards and regulations, information campaigns, introduction of economic and other incentives, which governments might implement in order to accelerate the uptake of low consumption/low emission vehicles on the market, were debated. The different entities which may play a role were considered, including central and local administrations, car manufacturers and users, fuel suppliers, population at large and R & D organisations. The experts explored and assessed the advantages and drawbacks of available options, their possible interactions and trade-offs.

The experts recalled that the increase in overall fuel consumption by automobiles in the 1980s stemmed mainly from: (i) the worldwide increase in the total automobile fleet, (ii) increase in mileage per vehicle, (iii) increased driving in congested urban areas, and (iv) the increasing power of many models presented on the market. They noted that the underlying factors which gave rise to these trends were the following:

a) The low level of fuel prices, and thus the small share this represents in the total cost of car ownership, fails at present to provide the necessary incentive for fuel efficiency both when the consumer chooses and buys the car and during its subsequent use. This has been enhanced by an impression by the general public of secure and abundant, cheap and stable oil supply.

b) The general increase in incomes for the majority of the population in OECD countries (and to some extent also in other regions of the world) which tends to make drivers less and less concerned by fuel economy.

c) The tendency of the general public to regard the automobile as a priority acquisition, and a trend in consumer preference and commercial publicity towards high-performance models.

d) The hesitation and inaction of many governments in taking adequate measures to offset these trends through the use of the economic and regulatory instruments at their disposal because of the general policies of reduced intervention in the market.

e) The low level of awareness by many entities of the full environmental consequences of automobile use, and the resulting tendency to attribute a low priority to achieving reduced automobile pollution levels through low fuel consumption.

The measures which are available to governments in order to counteract these influences include economic instruments on one hand, and regulations, standards and information programmes on the other. Amongst the most relevant economic instruments are:

(i) Fuel taxes: Some experts maintained that this has a crucial bearing on energy efficiency and conservation in general, which has already been discussed in many other fora on energy and environmental policies. Taxes could be applied as overall "emission charges", covering conventional pollutants and greenhouse gases. They should lead, if possible, to similar final prices of fuels in OECD countries, in order to minimise economic distortions and induce the desired convergence of policies. In Europe, for instance, the opinion of various experts was that, in order to be meaningful, the level of fuel taxation should be at about the same level as the existing average fuel price before tax. For North America, some experts thought that efforts should be made to approach general OECD fuel price levels. It was suggested that the revenue arising from such taxes could be directed not only towards the development and implementation of energy efficiency and pollution prevention measures at national level, but also internationally (i.e. funds for assisting technology development and deployment both in OECD and non-OECD countries). Other experts expressed pessimism, however, on the effectiveness of fuel pricing mechanisms, and noted that significant increases in gasoline prices might result in only modest gains in fuel efficiency trends. They felt that more dramatic improvements would require additional regulatory and other incentives.

(ii) Vehicle taxes: Many experts suggested that vehicle taxes, in general, should be based on fuel consumption, as it integrates in itself all incentives concerning emissions (greenhouse gases in particular) and energy conservation. They proposed that the policy approach should be to apply a much higher level of taxation on vehicles with high fuel consumption than on vehicles with low fuel consumption so that the car fleet can progressively become more fuel efficient. Such taxes should be applied at the time of purchasing the car, on the manufacturer and the consumer separately in order to induce the desired incentive on both. Additionally, these taxes might be imposed annually on the user, such as the "stamp" system already applied in various OECD Member countries.

(iii) Pricing and taxation systems for road use and access: tolls, urban parking, credits: Some experts proposed the imposition of pricing and tax systems, proportional to fuel consumption, in some specific sectors where these are not yet applied, and where they could generate powerful incentives. Examples proposed were: highway and bridge tolls; urban parking fees; credits for car purchase and insurance. In doing so, it was suggested that vehicles be visibly classified in relation to their fuel consumption (for instance, high, medium, low), identified by a stamp on the car (for instance red, yellow, green). It was pointed out that such a system of charges might create, through its repetitiveness, a psychological incentive much higher than its purely financial impact, and would as a result increase environmental awareness and motivation of drivers and general public.

Additionally, in the broad spectrum of regulations, standards and behavioral instruments, the attention of the expert panel focused on the following measures:

(i) Speed limits: Several experts recommended that existing speed limits should be strictly applied, especially in those countries where they are poorly respected. They further felt that speed limits should be identical for all automobiles (as is now the case in almost all OECD Member countries) for basic safety reasons and in order to discourage the choice of more powerful cars among the public. Experts recognised that there have been in a few countries some resistance to the concept of uniform or even restricted speed limits on major highways. However, most experts considered that the adoption of similar speed limits in all OECD countries would offer substantial advantages in terms of traffic management, safety, trade and tourism and suggested that a standard value for automobiles on highways, for instance, might be between 110 and 120 km/h with lower values for other roads depending upon their characteristics. This was seen by the same experts as a crucial policy issue regarding energy conservation, pollution prevention and safety as well as a basic measure of international solidarity for environmental protection which should be considered by OECD as well as non-OECD Member countries.

(ii) Regulatory (or agreed) targets for fuel economy: Experts noted that while fuel efficiency standards were successfully implemented in a few OECD Member countries in the mid-

1970s and achieved significant reductions in fuel consumption, there was a tendency in the mid-1980s, with low oil prices, to delay or even abandon such efforts. Some experts suggested that if governments wish to see further substantial improvements in fuel economy in the 1990s, they should aim now to establish and enforce the necessary targets, keeping in mind that potential fuel economy gains will, in general, be greater on powerful automobile models than on economical ones. Development of appropriate fuel efficiency standards was considered by these experts to be an urgent task. They pointed out that if this procedure is to be effective, the targets should be sufficiently ambitious, while innovation should be encouraged and market forces activated through competition. Several experts noted that whereas the protection of the environment is an accepted objective of business and industry, the identification by governments of clear and stable targets for fuel economy and emissions, which are technologically feasible and resource efficient with due attention to flexibility and timing, are the key ingredients for success.

(iii) Greenhouse gas emissions: The issue of greenhouse gas emissions was also discussed, and concerns about the impact of the increasing emissions of CO_2 from automobiles on possible climate change were emphasised. The amounts of CO_2 emitted by the combustion of a particular fuel are proportional to the quantity of fuel consumed by the car, and are thus strongly influenced by fuel efficiency. The experts thought that fuel efficiency standards were broadly relevant to the reduction of CO_2 and other greenhouse gas emissions. Experts referred to ongoing work in national and international organisations aimed at defining possible response strategies for the reduction of greenhouse gas emissions. In considering technological options which make use of different fuels, or in proposing a regulatory action which might result in a change in the type of fuel used, experts recognised the need for careful and detailed analysis of the influence of a particular option on CO_2 emissions, based on the full fuel cycle. For other greenhouse gases (e.g. N_2O) and the "indirect" greenhouse gases (NO_x, CO), experts considered that specific limits and targets should be designed when necessary, based on the best available information on the relative impact of these gases on climate change.

(iv) International standard test cycle: Test cycles for automobile fuel consumption received considerable attention and there was a fair consensus that existing tests are not sufficiently reliable in general. It was stated that at present, these tend to reflect "ideal" operating conditions quite different from those practised by the average driver in the range of traffic conditions normally encountered, with cold start, "average" car maintenance and typical driver behaviour. Some experts judged that the results of some tests could deviate by up to 30 per cent below the average values for fuel consumption obtained in actual driving conditions. They emphasised the fact that in addition to the difficulties this imposes on the selection of cost-effective technologies discussed earlier, such "optimistic" tests were detrimental to the correct information of consumers and to national policies and efforts to improve fuel efficiency. The development of a more reliable and internationally accepted test procedure was urgently advised by several experts, who considered that such a standard test could become the base reference for establishing government incentives, and for ensuring fair competition within industry.

(v) Inspection and maintenance: Experts noted that inspection and maintenance are necessary procedures which are already legal requirements in a number of OECD Member countries for pollution control, energy conservation and safety reasons, and should be considered for adoption by all Member countries. Experts noted that poor maintenance was a cause of greatly increased atmospheric emissions from automobiles fitted with otherwise-effective pollution abatement technologies, often resulting in performance several times worse than that of previously available, well-maintained cars without such equipment. There were experts who stressed the desirability for the newly industrialising countries, in particular, to initiate effective procedures for inspection and maintenance of road vehicles. This was seen as providing an effective means to these countries for monitoring and controlling the rapidly increasing atmospheric pollution which may be an inevitable result of their rapid economic growth.

(vi) Consumer information and environmental motivation: Within the framework of policies now being proposed in OECD Member countries on "environmental labelling" and the "green consumer", experts were of the opinion that the automobile can play an important role in view of its considerable impact on the environment and its share in consumer budgets. At present, consumers are often guided by commercial information and publicity alone, which tend to induce them to choose more powerful and high-performance models. The market is in fact rapidly growing in this direction. It was judged urgent by many experts to reverse this trend which could be environmentally detrimental, and obviously in conflict with policies being considered in many OECD Member countries and internationally to increase fuel efficiency. Experts urged that consumers should be provided with adequate general information on the influence of the automobile on the environment, including the economic and social advantages of reducing fuel consumption and pollution. Neutral and accurate information should be given on the fuel consumption and emission characteristics of automobiles available on the market, established according to real driving conditions, so that consumers are able to make a rational and responsible choice. National and international information campaigns should be carried out through advertising and other means of information dissemination, while applying, in parallel, adequate economic and regulatory incentives.

(vii) Driver education: The discussion here concentrated on the need to make car users aware of the impact which their driving habits and behaviour can have on safety, motoring costs and the environment (emissions levels, noise). Some experts felt that drivers should also be encouraged to integrate the use of public transport with the use of their own cars in urban, suburban and inter-city travelling, and that such driver education should be carried out during the initiation of driving instruction, and thereafter on a continuing basis through the media and other means. There were, however, mixed opinions among the experts on the ultimate efficacy of these actions unless they are reinforced by appropriate regulations and economic incentives.

5. NEED FOR AN INTEGRATED STRATEGY

The experts appraised the issues raised during the Panel meeting which have particular relevance to the development of integrated and co-ordinated policies and strategies for the deployment of low-consumption and low-emission automobiles. They recognised that the systemic character of the transport sector called for the convergence of objectives and solutions in energy, transport, environment and regional planning. The great diversity and international nature of the products, services, suppliers and consumers encompassed by this sector requires the setting of clear and co-ordinated targets which may be implemented simultaneously and equitably in all OECD Member countries. Experts were careful to note that the great diversity of interests involved will in general preclude the designation of a single theme or focus for such policy and strategy formulation. They generally agreed, however, on the desirability of defining an integrated framework within which the evolution of coherent sectoral policies and strategies could take place, and discussed the following guiding elements for such a framework.

Many experts felt that an integrated strategy should take account of accelerating changes in the ways that automobiles are produced and used. Developments in design and manufacturing technology reflect demand-side transitions in the OECD Member countries such as rapid growth in car ownership, changes in buyers' preferences and users' behaviour. All experts anticipated a rapid increase in car ownership in the newly industrialising countries, in Central and Eastern Europe and in the USSR. There was considerable agreement that this increase would largely be met by designs and technologies now available or under development in OECD Member countries. National and OECD policies and strategies should therefore be developed with a clear awareness of their likely imitative spill-overs and ultimate global impact.

The restrictions being imposed on the use of automobiles in many cities worldwide were seen by experts as already strongly increasing the demand for intermodal transport options. Innovations in urban mass transit systems allowing for smaller scale, greater flexibility, and park-and-ride linkages will define the context within which low consumption and low emission automobiles are utilised in urban areas. Experts stressed the need for policies which encourage optimum intermodal utilisation of rail, road, air and rapid transit infrastructures in high-traffic density regions and corridors. They recommended the detailed analysis of consumer requirements and aspirations, and ways in which these might in future be addressed and influenced through regulation, advertising, information, and other means.

Technological innovation was viewed by experts as essential to the increased deployment of less polluting and more fuel-efficient automobiles. Public and private development efforts will determine which technologies become available, and when. Experts noted that new concepts in automobile propulsion which were intended to meet the needs of particular groups of vehicle users may soon find wider markets in response to efforts to reduce vehicle pollution. The example of electric vehicles and their anticipated use in areas with serious air quality problems, such as the Los Angeles basin, was cited as a case where regional policies are providing a strong impetus to technological innovation. Many experts thought that immediate consideration should be given to identifying niche markets for the expanded use of electric vehicles and other innovative solutions.

Experts agreed that much of the vigorous and desirable competition which presently exists between manufacturers is centred on the quest for technological innovation. They noted that in parallel with this competition, there are useful roles for R&D collaboration among automobile and oil companies on technological options that have not yet reached the marketplace. Such collaboration may also involve national laboratories and universities. Subjects of particular relevance were thought to be: (i) new propulsion concepts such as electric hybrids, modular propulsion systems, advanced fuel cells; (ii) new gasoline formulations, improved liquid fuels and petroleum substitutes; (iii) new materials, permitting advanced designs for engines and car bodies; (iv) advanced electronics for real-time engine control and optimisation of the entire drive train; (v) highly efficient and flexible transmission systems; (vi) basic phenomena, including combustion science, catalysis, driver behaviour. Continued support by governments and international bodies for collaborative R&D in these areas was strongly endorsed. Experts further noted that a number of governments have already integrated such actions into national policies aimed at improving industrial competitivity, and urged the expansion of such efforts. Experts generally underlined the value of information exchange on policies, programmes, trends, results and experiences, and encouraged entities from all relevant sectors to examine the adequacy of existing efforts with a view to establishing improved procedures and mechanisms as required.

The experts concluded their discussions by recognising that policies and strategies aimed at reducing the present accelerating trend in automobile energy consumption through substantial increases in efficiency will have widespread benefits for the environment, energy security, balance of payments and sustainable socio-economic development. They noted that actions which originate from the suggested strategy framework will in any event be beneficial for all sectors even if some of the anticipated environmental and energy problems turn out to be less acute than presently foreseen. They felt that the creation of regulatory and economic frameworks and incentives which are stable, equitable and harmonised internationally would enable the automobile industry to better plan for future requirements. As a consequence, the industry would be in a better position to commit the resources required to achieve continuously improved levels in fuel consumption and emissions, in pace with technological progress. The resulting orderly change in automobile conception and design would provide the best assurance for the steady development of the industry, and would ensure that it can adapt and adjust in a planned way to the changing market place. Such evolution is vital to meet the crucial challenges of sustainable development and environment protection in this era of rapid global change.

PART II

OVERALL FACTORS INFLUENCING

FUEL CONSUMPTION AND EMISSIONS REDUCTION

POLICY OPTIONS TO ENCOURAGE LOW EMISSION/LOW FUEL CONSUMPTION VEHICLES

Deborah Lynn Bleviss

International Institute for Energy Conservation

Technologies have been developed and will continue to be developed that can reduce the fuel consumption of a light vehicle (automobile or light truck) significantly below that of today's standard light vehicle. But today's market forces provide little incentive to accelerate the introduction of these technologies. Oil prices are quite low, thereby offering little reason for manufacturers to pursue fuel economy. Moreover, at today's level of fuel efficiency for new cars of roughly 30 miles per gallon or mpg (7.8 l/100 km), fuel prices account for a relative small fraction of the overall daily costs of owning and operating a vehicle. Hence, even if fuel prices were to rise, the effect on consumers' pocketbooks would not be nearly as noticeable as when cars had much lower fuel economy.

Nevertheless, there are very valid and important reasons to pursue increased fuel economy in light vehicles. Most recently, the dominating reason has been environmental. Reducing a vehicle's fuel consumption reduces its carbon dioxide emissions--a major greenhouse gas--and usually reduces its other noxious emissions as well (e.g. carbon monoxide, nitrogen oxides, hydrocarbons). Without the market to spur increases in fuel economy, national governments must provide the necessary incentives instead.

The most effective way for national governments to do this is in a united manner. This will lessen the complaints of anti-competitiveness that would otherwise inevitably arise from automakers in one country that are being asked to improve the fuel economy of their vehicles while competing automakers in another country are not. It is important that the OECD countries reach mutual agreement on the levels of fuel economy that will be aspired to in their new light vehicles by a given date. In addition, agreement should be reached on the general types of mechanisms that will be employed to reach these targeted levels of fuel economy. But the specifics of designing the mechanisms should be left to each country since cultural variations and differing relationships between government and industry among the OECD countries require different approaches.

An aggressive but achievable goal for fuel economy is an average level of 45 mpg (5.2 l/100 km) for new automobiles and 35 mpg (6.7 l/100 km) for new light trucks by the turn of the century. To reach such a goal, a three-pronged governmental program will have to be initiated offering: (a) incentives to manufacturers to produce fuel efficient vehicles; (b) incentives to consumers to purchase such vehicles; and (c) incentives to pursue advanced research and development in fuel efficient technologies.

Incentives to Manufacturers

In 1975, the U.S. established the first set of fuel economy requirements for all manufacturers selling cars and light trucks in that country. That program was quite successful at bringing the level of fuel economy of light vehicles sold in the U.S. up to the level of other OECD nations. Following the creation of the U.S. targets, both Japan and most European countries with automaking capability established their own sets of targets, which were significantly less aggressive than the American standards.

Experience with these programs should form the basis for establishing new target programs for manufacturers. In the U.S., the long history of mistrust between government and industry meant that the targets had to be made mandatory in order to gain compliance. Since conditions between the government and the industry remain approximately the same today, this same approach should probably be maintained for a new set of targets. Nevertheless, there are some aspects of the old targets that should be changed. For example, the last set of targets established the same average fuel economy requirement for all companies, regardless of the mix of vehicles they sold. Hence, those companies selling predominantly small cars (mainly Japanese companies) had a much easier time meeting the standards than those selling both large and small cars (predominantly American companies). This inequity could be corrected in a new set of standards by requiring all companies to achieve the same percentage of average fleetwide fuel economy improvement.

In contrast to the U.S., the targets established by Japan were voluntary in nature; moreover, they were set by weight-class. With the long history of government and industry working together, the structure of these standards did induce the companies to seek to meet the targets. Hence, the same type of structure should be considered for a new set of targets, with a few exceptions. First, light trucks were not included in the first set of targets, even though they account for a significant fraction of sales in Japan. Consideration should be given to including them in a second set of targets. Second, the first set of targets only applied to Japanese manufacturers. Since these automakers presently dominate that market, such a structure makes sense. But if a decision is made to significantly open up the Japanese market to foreign manufacturers, they should be asked to meet future targets as well.

The European targets were also voluntary in nature, based again on a longstanding cooperative relationship between individual governments and their automotive industries. However, the European program was not as successful as the Japanese program. The European target programs were highly fragmented, varying in aggressiveness and form from country to country. And they were hard to monitor. These problems should be corrected in any future program. Recognizing the cooperative relationship between government and industry in Europe, voluntary targets could still be effective, but consideration should be given to establishing European-wide targets. In addition, the structure of the targets should be set such that they are easy to monitor whether compliance is occurring. Average fleet fuel economy targets, as well as weight-or size-class targets are two examples of structures that are easy to monitor. Finally, consideration should be given to including light trucks in the targets program.

Incentives for Consumers

Programs to encourage manufacturers to produce fuel efficient vehicles cannot be effective unless they are complemented by programs for consumers that foster their demand for such vehicles. Low fuel prices have certainly contributed to the present lack of interest by both consumers and manufacturers in fuel economy. Hence, consideration should be given by governments to increasing fuel prices. However, for the reasons listed previously, increased fuel prices are unlikely to stimulate much consumer demand for highly fuel efficient vehicles. Rather, they are likely to discourage demand for inefficient vehicles.

An option for directly encouraging consumers to demand fuel economy is through the price of the vehicle itself. In a way, this can be viewed as reflecting the lifetime cost of fuel consumption in the purchase price of the vehicle. Most OECD countries have some type of ownership fee or tax assessed on vehicles that is often based on weight, engine displacement, or engine horsepower. Hence, indirectly, the purchase of fuel efficient vehicles is encouraged. The U.S., on the other hand, has established a direct tax on fuel inefficiency called the gas guzzler tax. Amending existing ownership fees throughout the OECD to base them directly on fuel economy would strengthen the incentive for consumers to buy fuel efficient vehicles.

To encourage consumers to purchase vehicles at the top end of the efficiency spectrum, many have suggested a rebate or tax credit. Bills in the U.S. Congress presently offer a range of ways to structure such an incentive. Many suggest such an incentive should be given to the vehicles achieving high fuel economy within each size class, so as to promote the development not only of fuel efficient small cars, but large cars as well.

Research and Development

In order for fuel efficient vehicles to be manufactured, enhanced research and development efforts need to be stimulated. The way such stimulation should occur depends heavily on the country involved.

Japan's automakers and government have had a long history of working together to achieve certain technical goals. In that country, then, consideration should be given to making energy efficiency a high priority in this already ongoing cooperative effort. The importance of targeting fuel efficiency research in Japan cannot be underestimated, for Japanese automakers now largely set the standards in technology development. And their pursuit of technological progress has in turn stimulated their competitors to follow suit.

In Europe, individual governments and their automotive industry have also worked cooperatively on research and development. During the late 1970s and early 1980s, several European governments undertook cooperative research programs with their automotive industries to develop fuel efficient prototypes. Generally, these programs were quite successful, and thus consideration should be given to initiating new ones for the 1990s. However, several improvements to the past programs should be considered. The old programs were not oriented towards designing a vehicle that could be put into production; the new programs should. In addition, the old programs did not seek to ensure that other consumer requirements--such as comfort, adequate performance, safety, and acceptable emissions--were met when designing the prototype vehicles; the new programs should.

Unlike either Japan or Europe, the American automotive industry and the government do not have a good history of working together on research objectives. In fact, the only cooperative research program, called CARP, was dismantled shortly after it was created under pressure from the auto industry. Nevertheless, a new effort should be considered to establish a joint research program. Such a program should not only include the domestic automakers themselves, but also their suppliers and the small, innovative inventor businesses in the U.S. that have been behind much of the innovation in the automotive arena.

Conclusion

There are numerous reasons to encourage improved fuel economy in new light vehicles, not the least of which is growing concern about the threat of global warming. Since OECD countries account for about 80 percent of all the vehicles on the road today, they should take the lead in seeking to prompt these changes. The most effective way to do this is in a united manner, in which agreement is reached among all countries as to the level of fuel economy that should be achieved and the mechanisms that should be used to stimulate those fuel economy improvements.

Energy, Environment, and Travel.

*Lee Schipper, Ph. D.**

Lawrence Berkeley Laboratory

THE ENERGY AND ENVIRONMENT PROBLEM

The perception of the relationship between energy, environment, and travel has changed fundamentally over 20 years. After 1973, most observers felt that energy resource constraints, or higher costs, would be felt in the fuel tank, and slow the evolution of the economy. "Too little energy, too late" was the fear of many that provoked national and international programs to boost energy supplies, particularly those of oil from non-OPEC sources, as well as supplies of alternatives to oil. After weathering two major oil price increases, however, many began to realize that economic progress need not be limited by sheer resource constraints on energy, at least not for the foreseeable future. But the very success of throwing off the constraint of energy supplies has lead much of the world to view the energy-economy link differently, namely that total energy production and use is limited by growing environmental degradation. In short, the problem is now too much energy, too soon! The problem is the tailpipe, not the fuel tank. This is an important new paradigm for economists and planners. And this constraint will increasingly focus on the most visible aspect of energy use in the West, transportation.

Unfortunately, these problems are largely external and not internalized in present economic systems. That is, there are no economic feedback mechanisms stimulating consumers to use less polluting fuels. And the costs of cleaning up after the Alaskan or Los Angeles oil spills were not paid by those who had counted on the oil from the affected tankers, but by all those using oil products. But the gravity of environmental problems, exacerbated daily by new events or new scientific findings (ie., 1989 was another very warm year), means that environment is now permanently on the political agenda. Since economic policy instruments (ie., taxes) are still unpopular in the U.S., a major economic and political crisis looms as groups fight for the controls of policy instruments. The recent debate over the plan to clean air in Los Angeles is a small sample of what is to come. This is the environment in which the transportation and fuels industries must plan the approach to the next millenium.

Amidst all the concern over oil imports and higher prices, several fundamental changes in the way energy use used have progress in the U.S. and other countries. Energy use for production and distribution —manufacturing (including agriculture) and freight— has become relatively less important in Western societies, while energy use for personal services and private consumption— home comfort and convenience, personal mobility, and personal services— has increased in importance. This transition means that the small consumer's importance to total energy demand is increasing, while the large manufacturer's is decreasing.

These changes transform the energy market to one of many small consumers. But the fact that the consumers are dispersed means that policy tools that affect every consumer — prices, standards, certain prohibitions — must be called forth. Regulators in the U.S. have been afraid or unwilling to use taxes or other dispersed tools to affect consumers, relying instead on efficiency standards, while those in many other countries have traditionally relied more on taxes than on standards. Indeed, regulation in the U.S. has focussed on a few technologies that account for an increasing share of energy use: cars, air conditioners, other appliances, etc. Thus the structural change in energy markets exacerbates the energy/environment conflict for the transportation industry because transportation products are likely targets for regulation.

While these transformations have occurred, energy efficiency in the OECD (and the U.S.) has improved. Homes and buildings use 30% less heat per unit of area than they did in 1972, aircraft almost 50% less fuel per passenger-km, manufacturing about 25% less energy per unit of production, and cars in the U.S. about 30% less energy/mile. Significantly, however, cars in Europe and Japan have shown almost no net improvement, and the rate of improvement in the U.S. has slowed to a crawl. Indeed, efficiency improvements in many sectors have slowed. Meanwhile the total area heated (and cooled), the number of passengers flown, the output of industry, and the miles driven are rising. Combined with the slowdown in efficiency, these increased levels of activity mean more rapid growth in energy and oil use in the U.S. and other countries. While few see a major explosion in world oil prices, strong demand will put upward pressure on prices.

---

* The author is Staff Scientist with International Energy Studies, Energy Analysis Program, Applied Science Division, Lawrence Berkeley Laboratory These notes were written for the OECD/IEA Expert Panel on Low Consumption/Low Emissions Automobiles. Opinions expressed herein are solely those of the author.

The increase in energy use in transportation worldwide since 1973 is dramatic. While rapid increases in efficiency in the U.S. did limit the growth in travel related energy use, stagnation in efficiency elsewhere, combined with other factors, pushed up oil demand for transportation. Behind these increases lay several factors: greater car ownership, fueled by the maturation of the baby boom, splitting of households into smaller units each driving independently, more two-worker families, and more free time spent out of the home. These changes reduced the importance of mass transit, particularly for activities outside of normal commuting, boosted congestion in cities (increasing energy use). And more people are flying rather than driving, particularly in the U.S. In all, transportation is being forced to the forefront as the area of oil, energy and environmental growth, and therefore, the sector that public policy will deal with.

Now that the period of very high energy prices is (at least for the time being) behind us, energy per se is not driving consumer decisions. Instead, demography, work force composition, the housing market, etc., now more important to home and transportation energy use than traditional price and income factors. This is because these factors shape where live, how big the house are car will be, and how we connect living, playing, and working. Travel should be on this rise. This will lead to more pressure for solutions that reduce stresses between energy, environment, and travel. More generally, the reduction in these stresses will be important to rolling back the environmental constraints to human activity. The most important steps will focus on energy used by automobiles.

STEPS TO REDUCING AUTOMOBILE ENERGY USE.

There are many ways to reduce these stresses. All involve reducing energy use and pollution associated with travel. The alternatives are probably not exclusive, but they are not additive. And the value of each step is not independent of other steps. A major reduction in fuel intensity, for example, could reduce the impact of driving, measured per passenger mile, relative to that of busses, as well as reducting the cost per mile of driving. The challenge is to figure at each point how costly each marginal change would be, so as to order the steps. The discussion that follows reflects my own suspicions as to that order, but the "steps" overlap.

The first step is to attain a fleet average on the road approaching 40 MPG. It has been argued that if present day automobiles employed all the key features now present in at least some cars, the new car fleet in the U.S would attain new car fleet MPG to almost 40 MPG (test, corresponding to 33-35 MPG in the real world). If the entire U.S. auto and light truck personal vehicle fleet achieved 35MPG (as compared with present-day 19.5 MPG), oil use for this fleet would fall from 6 million barrels per day to about 3.5 mn barrels per day, cutting oil imports in the U.S. by about 1/3. The impact in Europe or Japan would be smaller because MPG is about 25MPG and total use is smaller that in the U.S.

The second step is less certain. A variety of high (>60MPG) prototype cars exist that have tested well. But these cars are somewhat smaller than the present fleets in the U.S. and Europe. The cars rely on new materials and new concepts. Most of the "new" elements in these cars will eventually find their way into market vehicles, but timing and cost are uncertain. Hence it is probably not realistic to simply pop these vehicles into our present models of travel and energy use. But these prototypes give evidence that the ultimate fuel economy goal for liquid fueled vehicles lies well above 50MPG. My own guess is that this step is *not* the next step.

Driving patterns is one reason. Better fuel economy is not a function of technology alone. Avoiding congestion (smart cars, better traffic controls, better trip planning) and coldstarts (better trip planning) are important to higher MPG. American driving, however, is characterized by more moving about in greater congestion, and more short trips. This feature of American lifestyles does give extra value to vehicles that are less polluting in congested areas, or that lose little our no fuel economy in congestion or cold start. Unfortunately, driving in every country is moving towards more congestion. Until congestion is reduced, either by reduced travel or through smart cars and highways, the truly high MPG cars may remain only prototypes, their true potential thwarted by long queues of traffic.

It is this consideration that has promoted interest in alternative fuels. First, society desires fuels the use of which in congested areas (ie., both in narrow city streets as well as in large air basins) is less harmful than using gasoline. The prospects are many: methanol, ethanol, compressed natural gas (CNG), liquid petroleum gas (LPG), reformulated gasoline, and, farther down the road, electricity and hydrogen. But these alternatives are *unlikely* to make a major contribution to a better environment because they cost more than gasoline and often require new infrastructure investments. Many alternatives have reduced performance (or range) relative to gasoline. All of these fuels can make a contribution, of course, but a careful reading of the literature suggests that reductions in intensity are *both* cheaper per mile driven than alternative fuels *and* a necessary first step if the performance and range of alternatives are to be increased.

Another step may be mass transit. Authorities in Los Angeles and other cities have pinned hopes on mass transit to provide a valve for reducing tension between travel and the environment. Unfortunately, mass transit shows little promise *under present conditions*: cheap fuel, zoning and tax laws that favor sprawl, cheap parking, few if any charges to enter roads or cities at congested times etc. And while the share of transit is higher in Europe or Japan than in the U.S., its role has been eroded constantly by the automobile. Finally, driving for commuting in the U.S., measured as per capita passenger kilometers travelled, has been stagnant in the 1970s and early 1980s, in part because transit has maintained a respectable share of commuting trips and travel. Instead, it is travel for family business and free time that has grown. This travel, which represents about 2/3 of auto travel, takes place in less orderly or regular patterns than commuting. and therefore is less easily substituted by bus or rail transit that takes fixed routes at fixed times, at least in American cities. Thus it is hard to imagine transit relieving driving of all but a small share of miles. Indeed, a doubling of transit passenger-km, from just under 3% (excluding school busses, which account for another 2-3%) to 6% would represent a great victory for transit, yet leave approximately 85% of all travel, and 97% of present automobile travel, in the automobile. This pessimistic assessment means that one highly regarded substitute for auto travel can only make a marginal contribution to relieving travel/environmental pressures. And should lavish subsidies encourage districts to run busses or trains more often than not, empty seats could push the energy use of transit to levels comparable to that of more efficient cars.

Finally, what are the prospects of a different relationship between man (or woman) and the wheel? Above we argued that current trends point towards more driving, indeed, in more congested areas. And there is little interest in changing the relationship between where we live, where we play, and where we work, which impact how much we go. Indeed, inspection of data from periods when gasoline availability was restricted, or gasoline was very expensive (1974, 1979) show that people drove less during times of immediate crisis, then backed off towards smaller, and then more efficient cars. Thus I conclude that the prospects for reduced driving as grim, at least relative to the other steps covered previously.

What to Do?

My bleak assessment of the prospects for improved energy efficiency in transportation may only reflect the difficult American situation, where gasoline taxes are still unwelcome and politicians keep looking for the technical fix as a free lunch. But recent experiments with city tolls (Oslo), higher gasoline taxes (Norway), an explicitly CO_2 tax (Sweden), and talk of higher gasoline prices in urban areas (Stockholm) suggests that some countries are prepared to come to grips with the problem.

I believe that a combination of policy measures could accelerate the cleansing of the air and a reduction in energy use for automobiles. First, countries must decide what the true cost of travel and using gasoline is and price these commodities accordingly. Second, countries must get tough with automobiles at the point of purchase, designing standards or taxes that effectively promote fuel economy. This means reexamining company car policies that often reduce the cost of driving to nearly zero. Finally, countries, regions, and cities must re-examine policies that affect the patterns of settlement—where we work, live, and play— to see which policies have encouraged more sprawl and travel, and what yet unexecuted initiatives might reverse some of the trends that have increased energy use and pollution from travel.

TRAVEL AND ENERGY: CONFLICT OR COMPLIMENT?

PRESENTATION TO

OECD/IEA SEMINAR ON LOW-ENERGY/LOW-EMISSIONS CARS

ROME, FEBRUARY, 1990

Lee Schipper, Ph. D.

INTERNATIONAL ENERGY STUDIES, ENERGY ANALYSIS PROGRAM

APPLIED SCIENCE DIVISION, LAWRENCE BERKELEY LABORATORY

---

* Dr. Schipper, who is solely responsible for the opinions expressed herein, is Co-Leader of International Energy Studies, Lawrence Berkeley Laboratory. The author of "Coming in from the Cold: Energy Wise Housing from Sweden" and several other international analyses of energy use in Sweden and other countries, he received the Royal Medal from H.M. King Karl XVI Gustav for his research on Swedish and American Energy Issues. He has been a guest researcher at the Royal Academy of Sciences and VVS Tekniska Foerening, Stockholm, as well as at Group Planning, Shell International, London. He is also Senior Associate of Cambridge Energy Research Associates.

ENVIRONMENT: THE NEW ENERGY CONSTRAINT?

THE THREAT IS TOO MUCH ENERGY, TOO SOON; NOT TOO LITTLE, TOO LATE!

- LARGE SCALE LAND, WATER DISRUPTION FROM ENERGY PRODUCTION, TRADE
- LOCAL POLLUTION FROM FOSSIL FUELS, PARTICULARLY FROM TRANSPORT
- GLOBAL CLIMATE THREAT FROM GREENHOUSE GASES LINKED TO ENERGY SYSTEMS

FEW COUNTRIES INTERNALIZED CONCERNS INTO PRICES OR POLICIES YET

ENERGY DEMAND UNDERGOING FUNDAMENTAL STRUCTURAL CHANGES:

- RATE OF EFFICIENCY IMPROVEMENTS SLOWING MARKEDLY
- ENERGY DEMAND GROWTH IN OECD FOCUSED ON THE CONSUMER
- ENVIRONMENT EMERGING AS MAJOR GLOBAL CONSTRAINT

THESE CHANGES CHALLENGE ENERGY POLICY WHEN ENERGY IS CHEAP!

ENERGY DEMAND UNDERGOING FUNDAMENTAL STRUCTURAL CHANGES:

- RATE OF EFFICIENCY IMPROVEMENTS SLOWING MARKEDLY
- ENERGY DEMAND GROWTH IN OECD FOCUSED ON THE CONSUMER
- ENVIRONMENT EMERGING AS MAJOR GLOBAL CONSTRAINT

THESE CHANGES CHALLENGE ENERGY POLICY WHEN ENERGY IS CHEAP!

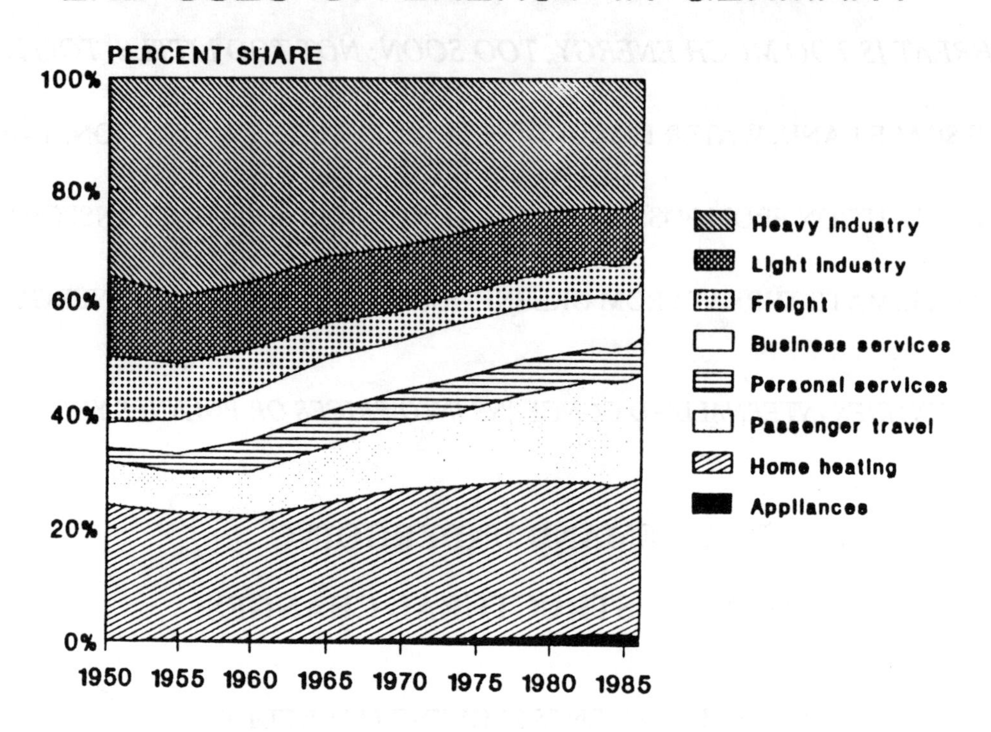

OECD COUNTRIES: ENERGY EFFICIENCY GAINED 1973 - 1988

PAST ACHIEVEMENTS IMPRESSIVE

• SPACE HEAT: NEW HOMES 30-50%% LESS LEAKY THAN IN 1973

N. AMERICA AND EUROPE SHOWED GAINS

• ELECTRIC APPLIANCES: NEW MODELS 20-40% LESS INTENSIVE THAN IN 1973

THE MARKET IS NOW WORLDWIDE

• AUTOMOBILES IN N. AMERICA, 30% LESS GAS GUZZLING THAN IN 1973

ALMOST NO IMPROVEMENT IN EUROPE, JAPAN!

• AIRCRAFT 40-70% MORE EFFICIENT THAN IN 1973, STILL IMPROVING

ORDERS FOR STATE-OF-ART AIRCRAFT RISING

• OECD INDUSTRIES REDUCED ENERGY INTENSITIES 20-35%

JAPAN LED

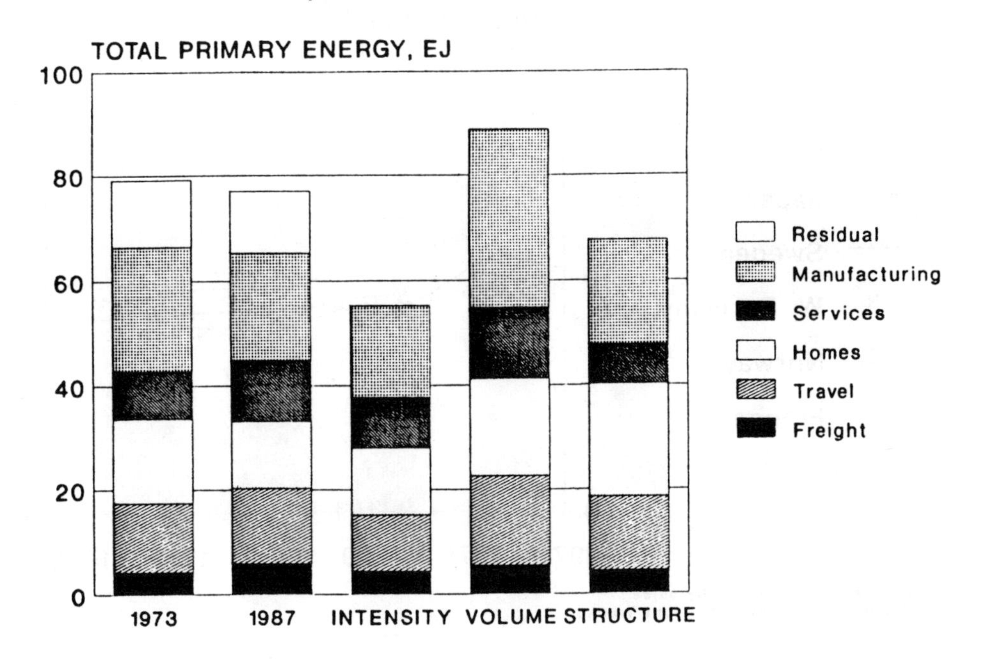

IS ENERGY CONSERVATION IN OECD BECOMING UNGLUED?

RISING OIL IMPORTS AND "TEMPERATURES" CAUSE CONCERN

- SAVINGS SINCE 1973 WERE AT LEAST 16 MB/DOE ALL FUEL, 12 MB/DOE OIL
- BEHAVIORAL/SHORT TERM COMPONENT REVERSING
- IMPROVEMENTS IN AUTOMOBILES SLOWING OR REVERSING
- APPLIANCE IMPROVEMENTS ALSO STAGNANT

ENERGY USE FOR TRAVEL UP EVERYWHERE

AUTOMOBILE FUEL ECONOMY
New Car Fleet Averages (Tests)

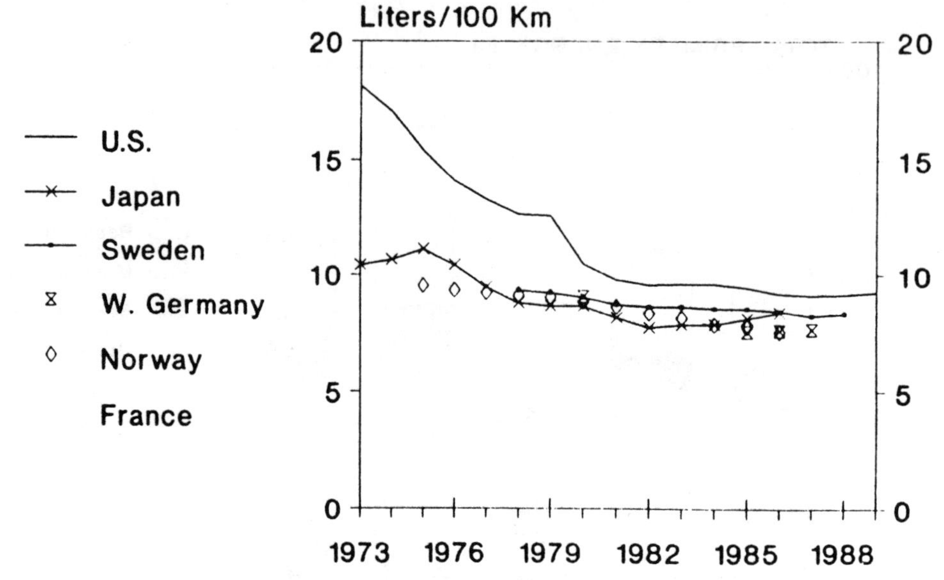

Figures weighted approx. by sales
US Figures (1989 est) incl. Light Trucks

AUTOMOBILES FUEL ECONOMY ON THE ROAD FLEET AVERAGES

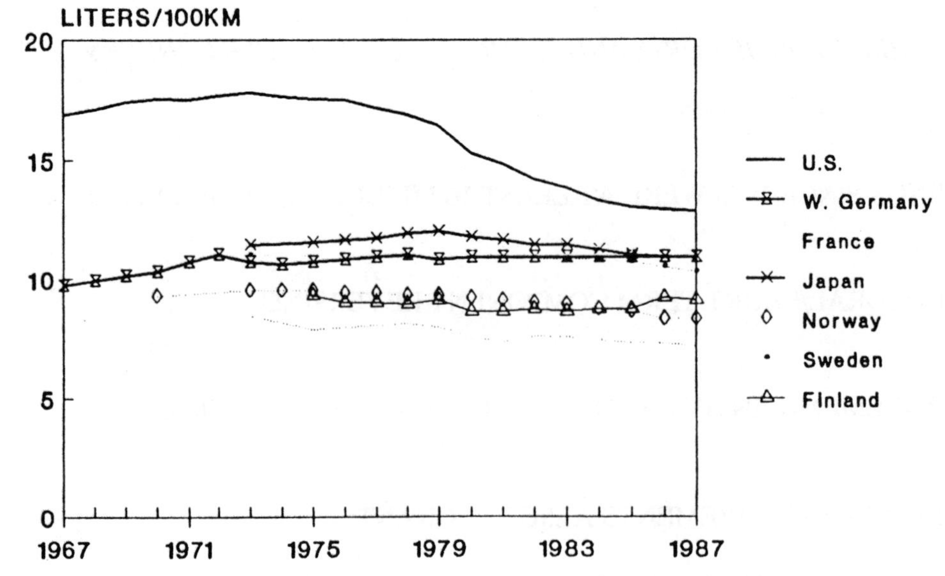

US, DOE; W. Germany, DIW; Norway, Esso
France, AFME; Japan, MITI; Sweden, TR
Finland: KTM

FROM PRODUCTION TO PLEASURE: THE CONSUMER*

PEOPLE WANT MORE CHEAP COMFORT AND MOBILITY

- HOUSE SIZE GREW 15% SINCE 1972, SPACE/CAPITA GREW 25%
- APPLIANCE OWNERSHIP "DOUBLED"; AUTO DISTANCE/CAPITA GREW
- TOTAL TRAVEL/CAPITA GREW 30% OR MORE, SHIFTED TO CAR AND RAIL
- SERVICE SECTOR EXPANDED AS CONSUMERS "OUT" MORE OFTEN
- FAMILY STRUCTURE EVOLVING TOWARDS GREATER ENERGY USE

PEOPLE SHARE OF ENERGY USE IS RISING

THESE TRENDS INCREASE ENERGY USE AND MAY COLLIDE WITH CLIMATE

* "Linking Energy Demand and Lifestyle". LBL project sponsored by EPRI and AGIP.

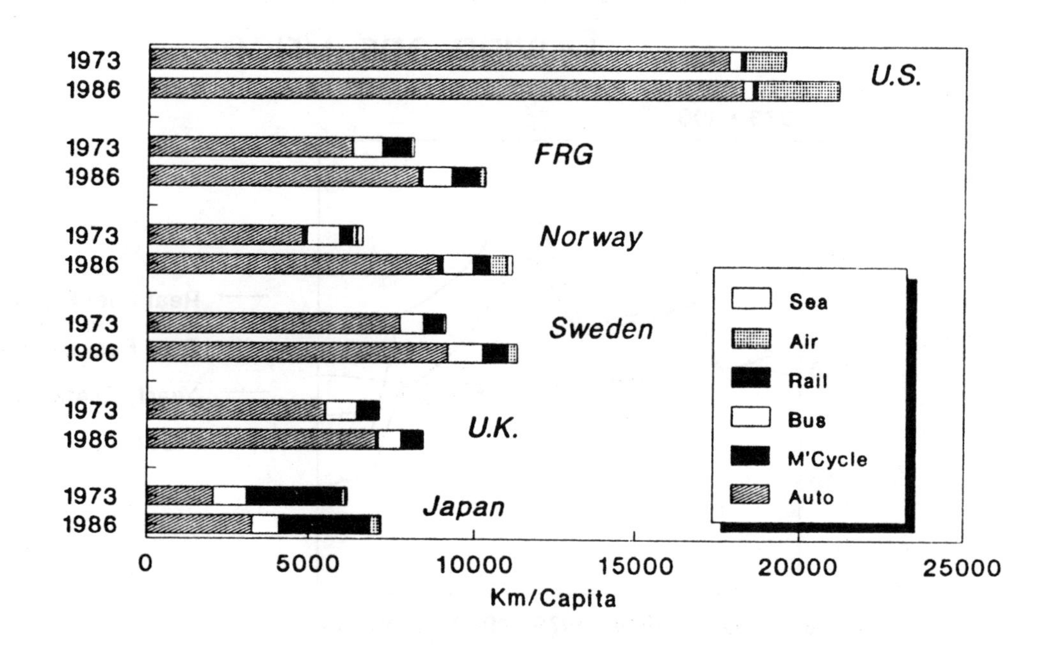

U.S., OECD TRANSPORTATION ENERGY USE 1970 - 1987

TRANSPORTATION KEY ENGINE OF GROWTH IN OIL USE

- VOLUMES DECLINED SLIGHTLY RELATIVE TO GDP
(VOLUME OF TRAVEL INCREASED REL TO GDP IN EUROPE)

- MODAL SHIFT INCREASED TRAVEL ENERGY
(SAME MODAL SHIFTS OCCURRED IN EUROPE, JAPAN)

- FUEL INTENSITY OF CARS, AIR DECLINED MARKEDLY, NOW STEADY
(CAR FUEL ECONOMY IN EUROPE, JAPAN CONSTANT SINCE 1973!!)

- FUEL INTENSITY OF TRUCKS DECLINED SLIGHTLY, HANDLING WORSENED

IS TRAVEL AND FREIGHT ACTIVITY SATURATED?

IS EFFICIENCY STAGNANT?

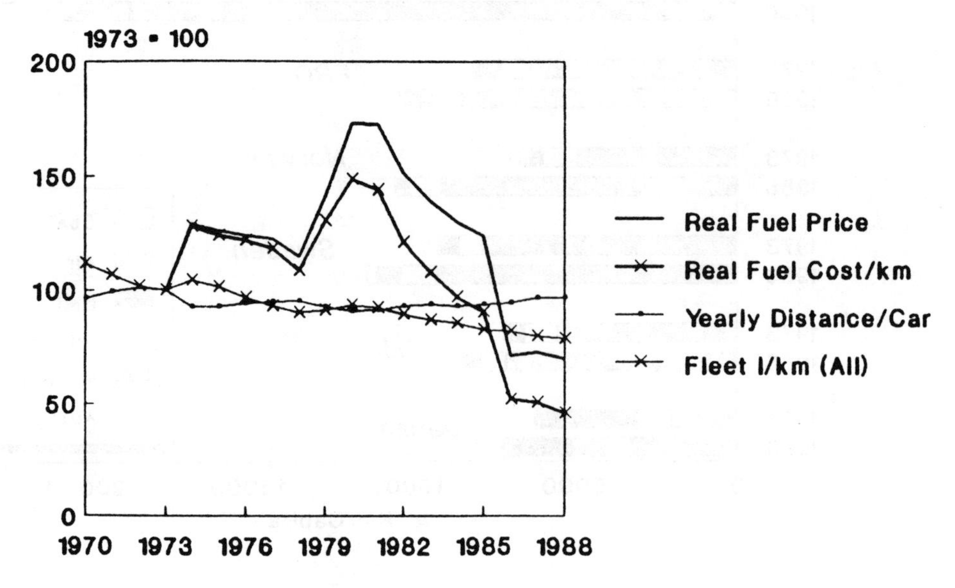

U.S. AUTO FUEL ECONOMY
Fleet and New Autos

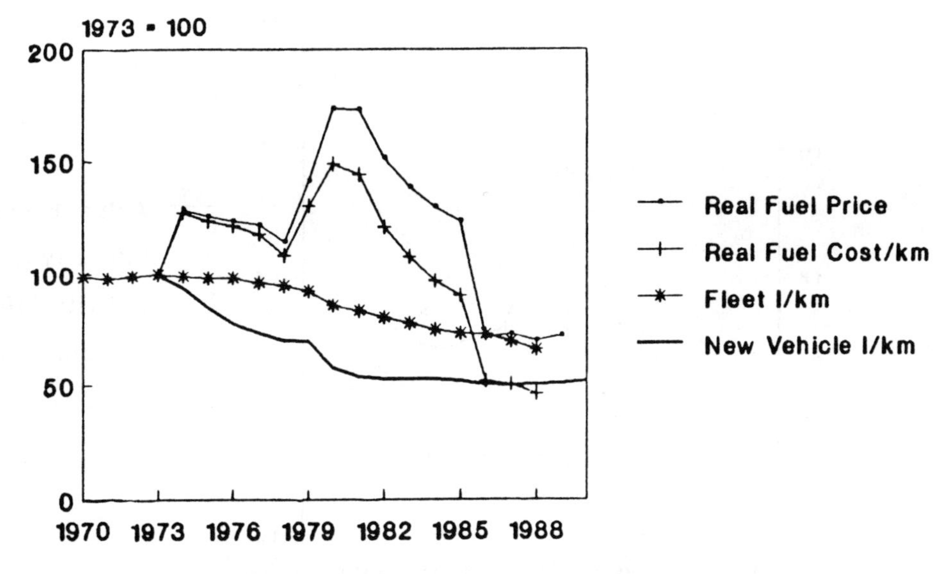

Includes Personal Light Trucks
Source: ORNL, ACEE, and IES

Passenger Transportation by Mode

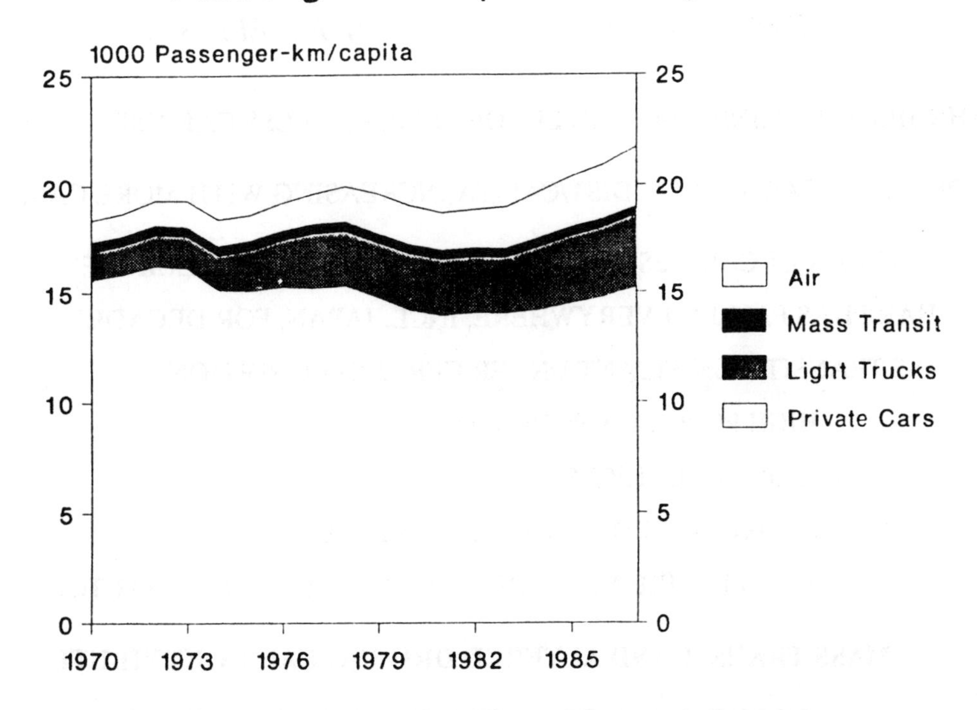

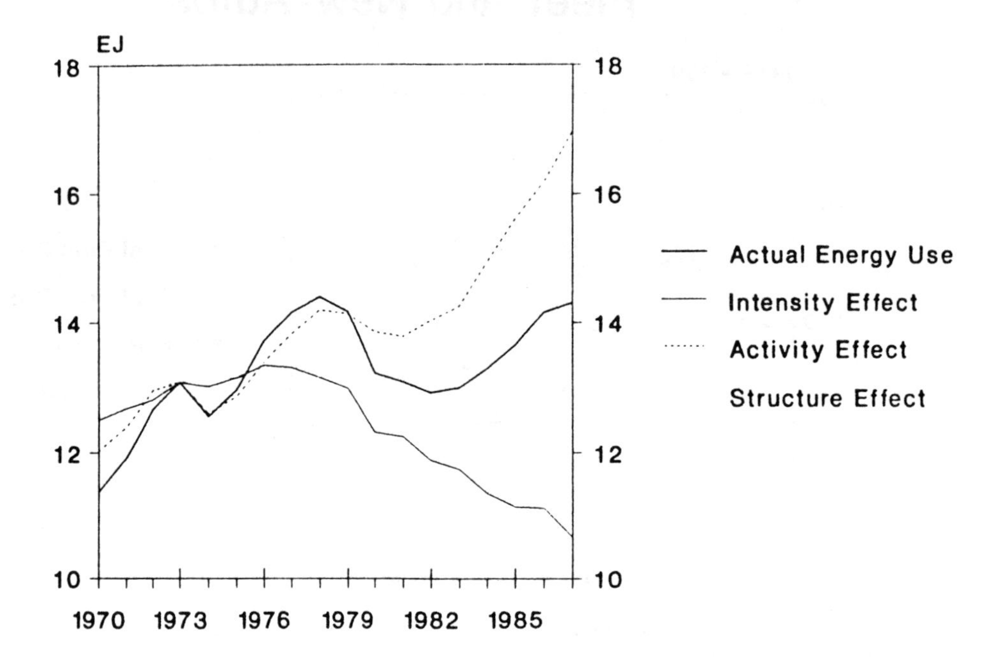

DRIVING AND MASS TRANSIT

HARD LOOK AT INTERNATIONAL EXPERIENCE

- HIGHER GASOLINE PRICES IN EUROPE LEAD TO LESS CAR USE
- DISTANCE/CAR STABLE, DIST/CAPITA INCREASING WITH MORE CARS
- MASS TRANSIT CARRIES LESS THAN 20% OF TRAFFIC IN EUROPE!

-- SHARE HAS FALLEN EVERYWHERE, INCL. JAPAN, FOR DECADES

- MASS TRANSIT SIGNIFICANT UNDER CERTAIN CONDITIONS:

  -- HIGH POPULATION DENSITY

  -- HIGH FUEL PRICES

  -- COORDINATED LAND USE CONTROLS

  -- TRAFFIC LIKE NEW YORK, LONDON, BANGKOK, OR TOKYO!

MASS TRANSIT AND ALTERED DRIVING CAN CONTRIBUTE,

IF PEOPLE ARE PREPARED TO BITE THE BULLET

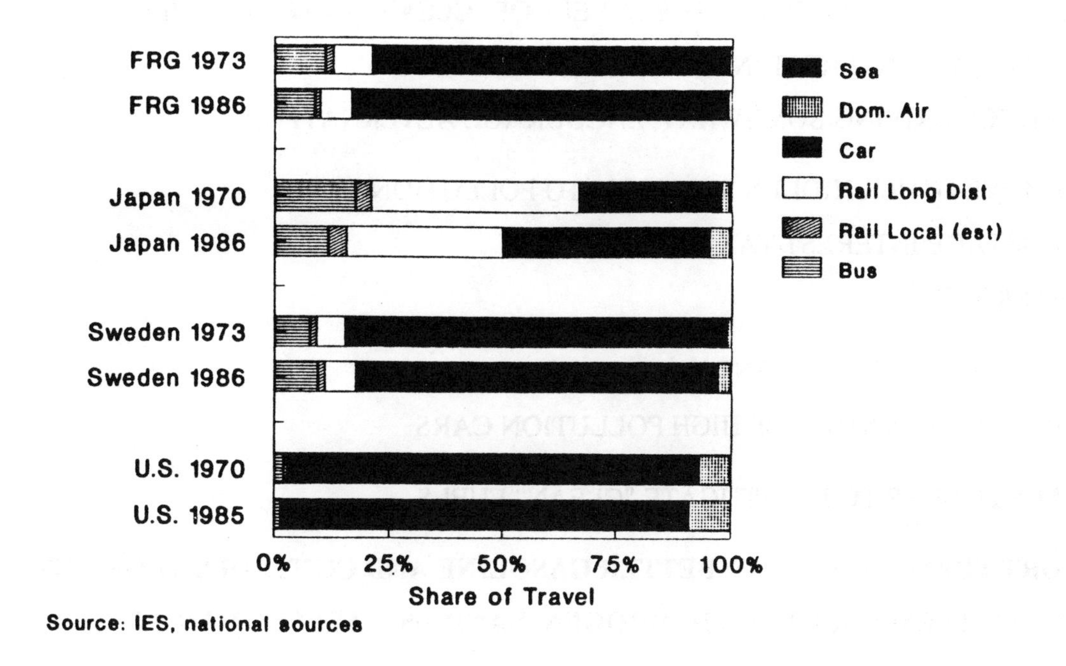

ALTERNATIVE FUELS OR ALTERNATIVE FOOLS?

ALTERNATIVES SIGNIFICANTLY MORE EXPENSIVE THAN GASOLINE

- FULL COST OF METHANOL, INCLUDING CO_2, HIGHER THAN GASOLINE
- ELECTRIC VEHICLES REPRESENT VERY HIGH COST FUEL, TOO
- HYDROGEN OFFERS FULL FLEXIBILITY OF GASOLINE, LOW POLLUTION
- ALL ALTERNATIVES REQUIRE SIGNIFICANT INFRASTRUCTURE INVESTMENTS

IF ALTERNATIVES NECESSARY ....

GASOLINE PRICES SHOULD BE RAISED TO LEVEL OF ALTERNATIVES

AN ALTERNATIVE PATH

MAKE THE ECONOMIC SYSTEM WORK FOR CLEAN AIR AND ENERGY EFFICIENCY

- TAX OFFENDING FUELS UP THE LEVELS OF "CLEAN" ALTERNATIVES

-- STIMULATE INTEREST IN HIGH MPG CHOICES

-- REDUCE DRIVING SOMEWHAT, ENCOURAGE ALTERNATIVES

- REMOVE OTHER HIDDEN SUBSIDIES TO POLLUTION

-- MORTAGE INTEREST TAX DEDUCTION?

-- PARKING?

- ENCOURAGE INVESTMENT IN CONTROL EQUIPMENT

-- HEAVIER TAXATION OF HIGH POLLUTION CARS

- TAKE LONGER TO INVESTIGATE "CLEAN" FUELS

SHORT TERM SOLUTIONS: BETTER GASOLINE AND CONTROLS, HIGH MPG

LONG TERM SOLUTION: HYDROGEN, LAND USE, TELECOMMUTING

THE ENVIRONMENTAL/ENERGY DILEMMA

THE "RIGHT" GREEN/CLEAN TECHNOLOGIES EXIST, BUT

- THE COSTS ARE MANAGEABLE, BUT FIRST COSTS LOOK LARGE

- THE CURRENT BUSINESS AND CONSUMER CLIMATE POINTS THE OTHER WAY

- TRADEOFFS AMONG ENVIRONMENTAL OPTIONS NOT ALWAYS CLEAR

- BIG ENERGY AND ENVIRONMENT SAVING POTENTIAL MUST BE STIMULATED

IF NO STEPS TAKEN, TRANSPORT, ENERGY, AND ENVIRONMENT IN CONFLICT

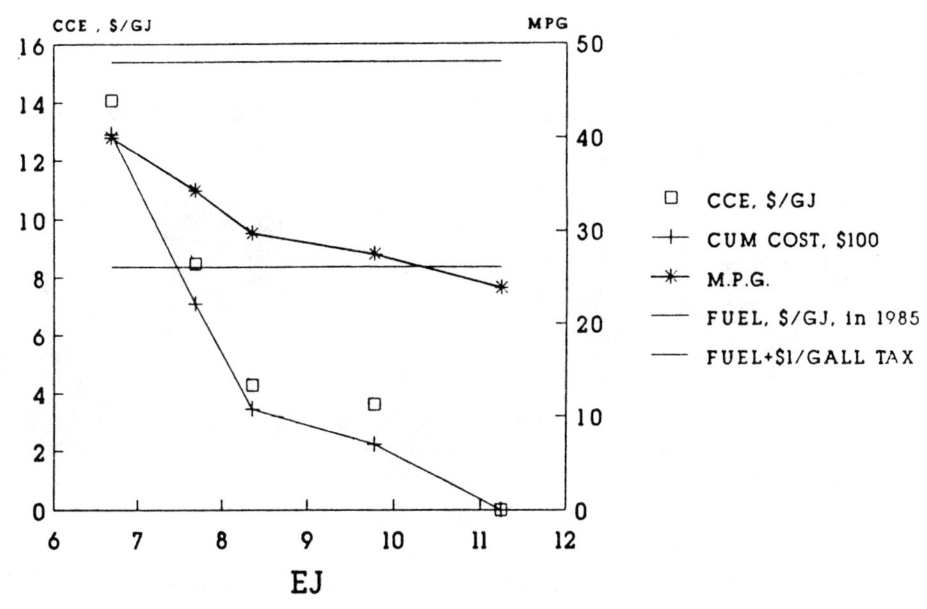
U.S. ENERGY DEMAND SCENARIOS
Cost of Saved Energy For New Cars
CCE , $/GJ
MPG
16
14
12
10
8
6
4
2
0
50
40
30
20
10
0
6
7
8
9
10
11
12
EJ
CCE, $/GJ
CUM COST, $100
M.P.G.
FUEL, $/GJ, in 1985
FUEL+$1/GALL TAX
6% Real Discount, 10 yrs, 1985 Prices
Technical Data: Duleep et el 1989

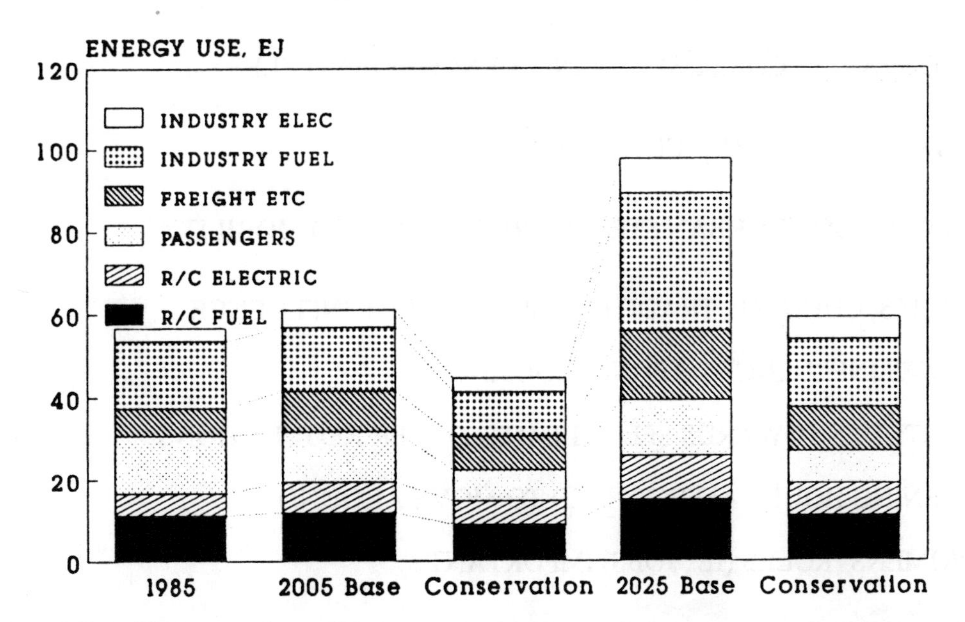
U.S. ENERGY DEMAND SCENARIOS
IMPACT OF MORE EFFICIENT USE
ENERGY USE, EJ
120
100
80
60
40
20
0
INDUSTRY ELEC
INDUSTRY FUEL
FREIGHT ETC
PASSENGERS
R/C ELECTRIC
R/C FUEL
1985
2005 Base
Conservation
2025 Base
Conservation
Source: IES
1 EJ = 1000PJ = 0.47mb/doe = 239MTOE

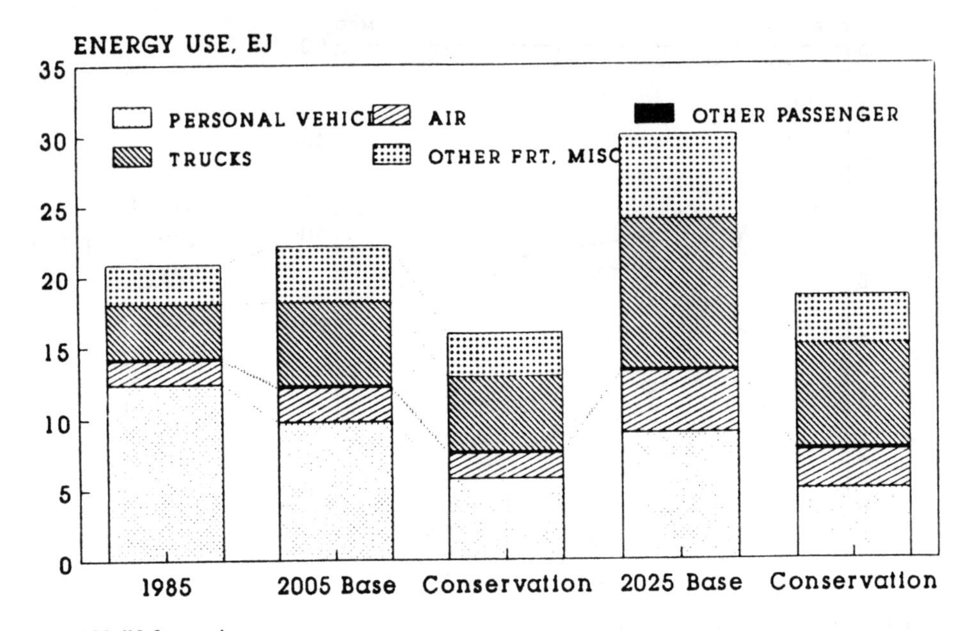

LONG TERM ISSUE: ENERGY AND LIFESTYLE

OTHER IMPORTANT CHANGES IN WHO WE ARE AND WHAT WE DO

- DEMOGRAPHIC COMPOSITION INFLUENCES LIFESTYLES
- AGEING WILL INFLUENCE ACTIVITIES
- CHOICES ABOUT URBAN STRUCTURE CRUCIAL TO MOBILITY
- POLICIES, CULTURE, TRADITIONS ALSO EXERT INFLUENCE

--- TAXES INFLUENCE DRIVING, HOUSING

-- ATTITUDES TOWARDS SPEED LIMITS, OTHER REGULATIONS

--- OPENING HOURS AFFECT NEED TO MOVE AROUND

--- WOMENS' ROLES (IE., JOBS) IMPORTANT

THESE SOCIAL FACTORS HAVE ENORMOUS INFLUENCE ON TRAVEL DEMAND

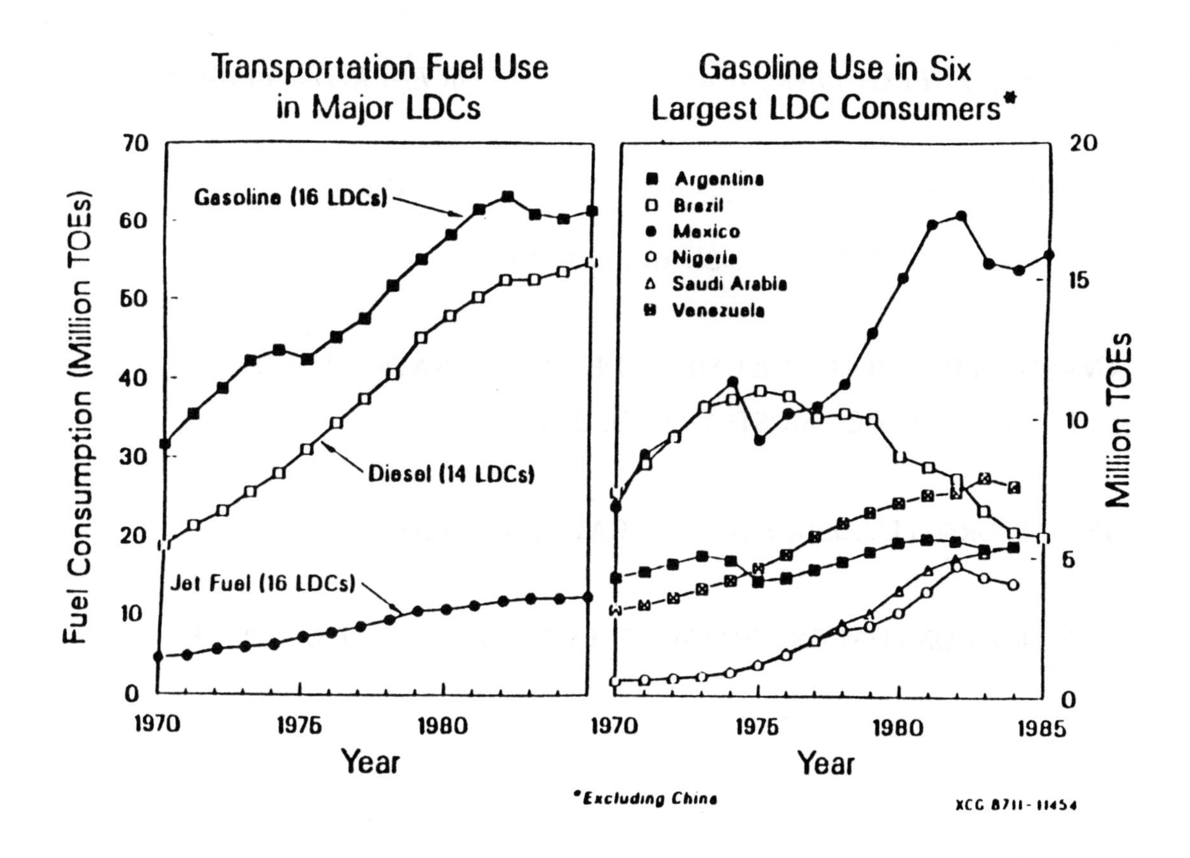

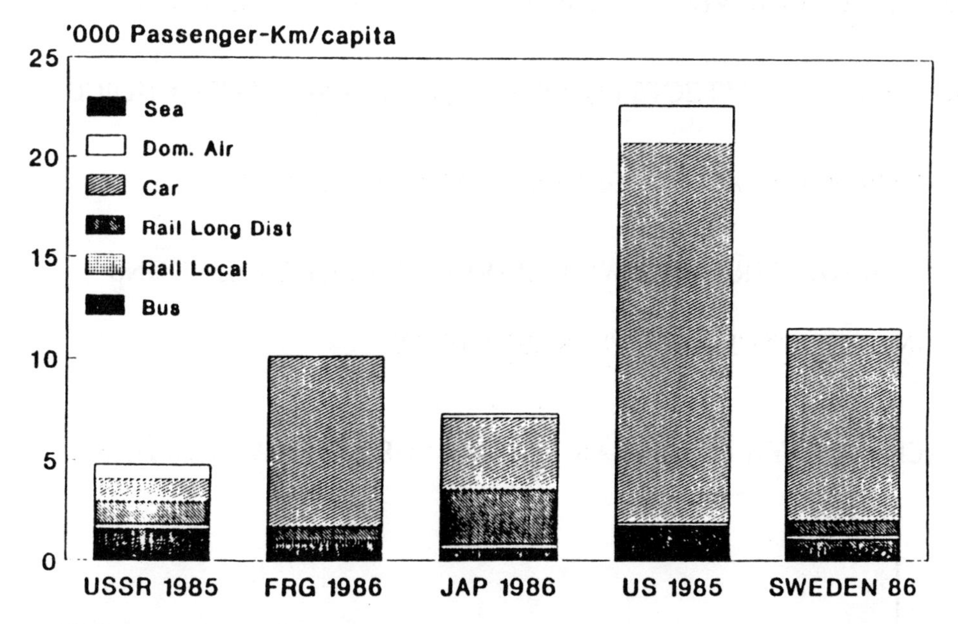

Source: IES/Lawrence Berkeley Lab

CONCLUSION: LONG TERM ISSUES

LIFESYLE-ENERGY LINK MEANS TROUBLE FOR ENERGY-CLIMATE LINK!

- ENERGY DEMAND: THE PRODUCER'S SHARE SHRINKS

  THE CONSUMER'S SHARE EXPANDS

- VARIABILITY FROM CONSUMER SIDE ± 15% OF NATIONAL TOTAL

  TOTAL VARIATION MOSTLY IN OIL DEMAND

- NO ASSURING EVIDENCE OF SATURATION OF TRAVEL

- IMPLICATIONS FOR ENVIRONMENT: GUARD AGAINST THE UPSIDE

AN ENVIRONMENTALLY CONSTRAINED ENERGY FUTURE?

GREATER DEMAND FACES GREATER CONSTRAINTS

- HIGHER ACTIVITY LEVELS -> LOWER IMPACT/UNIT OF ACTIVITY
- LOCAL POLLUTION AFFECTS EQUIPMENT, EFFICIENCY, FUEL CHOICE
- GLOBAL PROBLEMS AFFECT GREENHOUSE GAS EMITTERS
- INTERNATIONAL STRUGGLE WITH MARKETPLACE, REGULATIONS
- WAFFLING OVER "SCIENTIFIC" UNCERTAINTY

SCENARIO YIELDS BLEAK FUTURE FOR ENERGY POLITICS

CONCLUSIONS: ENERGY POLICIES AND OTHER POLICIES

SACRED COWS MAY HAVE TO BE MILKED

- ENERGY PRICES MUST REFLECT CONCERNS, NOT JUST COSTS
- "EFFICIENT" EFFICIENCY INTERVENTIONS NEEDED
- SOME INDIRECT POLICY OPTIONS:

  SPEED LIMITS (IN GERMANY)

  COMPANY CAR TAX TREATMENT (UK, SWEDEN, NORWAY)

  MORTGAGE INTEREST DEDUCTION (US)

  LAND USE AND TRAFFIC POLICIES

  GASOLINE TAXES (US!)

CONCLUSIONS: CONFLICT OR COMPLEMENT?

RESTRAINING AND CLEANING UP ENERGY DEMAND TO AVOID CONFLICT

- LIFESTYLES, NOT TECHNOLOGY, MAY BE A LIMITING FACTOR
- ROLE OF PRICES, POLICIES, PROPAGANDA STILL UNCERTAIN
- CONNECTING ACTORS (CONSUMER, PRODUCER, AUTHORITIES) KEY TO SAVINGS
- SPREADING HARD-, SOFT-WARE TO PLANNED ECONOMIES, 3rd WORLD IMPORTANT

CHALLENGE TO INDUSTRY: PRO-ACT OR RE-ACT? LEAD OR FOLLOW?

DEFINE COMPLEMENTARY PATHS FOR ENERGY, ENVIRONMENT, AND TRANSPORT

ENERGY EFFICIENCY IN ROAD TRANSPORTATION

Marcello Roma
Directorate General for Energy
Commission of the European Communities

A. "INEFFECTIVENESS" OF PAST TECHNOLOGICAL PROGRESS ON ACTUAL CAR USE AND FUEL CONSUMPTION

Definition of energy efficiency in the transportation sector

Energy efficiency in transportation can be considered as the relation between the energy consumption and the number of vehicles per km (or passengers per km). The calculation of this factor has to be done separately for private cars, buses and heavy vehicles because of their different unit consumption ratios.

Evolution of consumption, mobility and energy efficiency

Between 1980 and 1986, traffic (vehicles per km) of heavy vehicles increased more strongly (+25.4%) than that of private cars (+14.7%). During this period, global energy consumption increased by 14%. Consequently, there followed a smooth improvement in energy efficiency related to the mobility. Considering the breakdown of consumption of 60% for private cars and 40% for heavy vehicles, this gain in energy efficiency amount to 1.0% for the total period or 0.17% per year (due to technological progress). This disappointing result becomes even worse, turning into a net small decrease of energy efficiency if one considers passengers per km and tons per km, instead of vehicles per km. The main reason for this, arises from both the drop of the vehicle occupancy rate (-1.3% between 1980 and 1986) and of the loading rate of the heavy vehicles (-1.4% between 1980 and 1986).

Some explanations of the disparity between technical progress and real consumption

- The slow penetration of technological improvements in the vehicle fleet (in relation to its rate of replacement): The progress realised between 1978 and 1985 in terms of reduced consumption for new vehicles sold in Europe can be estimated according to the manufacturers :

  Reduction of 19% for gasoline cars (2.7% per year)
  " " 26% for diesel cars (3.7% per year)
  " " 20% for heavy vehicles (2.9% per year)

  With a replacement rate of the fleet of 10% per year, the energy efficiency improvement of new vehicles will after 10 years reach 27% whereas the average (theoretical) gain of the fleet is 15%.

- The gap between the consumption declared by the manufacturers and reality: In the Federal Republic of Germany, the average consumption data supplied by the DIW (Deutsche Institut für Wirtschaftsforschung) comes from tests executed by ADAC (Allgemeine Deutsche Automobile Club). This data differs greatly from that of the manufacturers. Thus, according to them, the average consumption decrease amounts to 19% for gasoline cars and 26% for diesel cars between 1978 and 1985. According to the DIW this decrease is in fact 13% and 18% respectively. (These assessments concern <u>all</u> cars sold in the Federal Republic).

- Traffic conditions: the increase of traffic movements over the past years can be essentially explained by the increase of urban traffic. Cars consume more energy in urban traffic conditions than in open country.

- Driving behaviour and the use of vehicle: user behaviour is essential. For example, the training organisations generally estimate that 10% of the potential for energy reduction can be reached thanks to good driver training. On the other hand, greater utilisation of the vehicle in urban conditions, leads to an over-consumption of energy (traffic jams, short length of journeys, non-stabilised engine running etc.) In this respect, an important saving potential would consist in optimising the engine characteristics for urban use. Some ideas are presented below.

B. BETTER COMPLIANCE OF THE CAR WITH REAL URBAN TRAFFIC

All aspects of the conception and design of modern automobiles are optimised for open road and highway driving. However these factors currently do not adapt for urban usage especially in terms of energy and pollution efficiency. Most surveys shows that about half of average car mileage is accomplished in urban areas. A fundamental inequality between an automobile and its real usage thus appears evident. This gap between conception and usage of the private vehicle may well be one of the causes of pollution in large cities. In the short term, efforts should aim to move the maximum of the torque curve to a lower r.p.m., a figure more appropriate to urban traffic conditions. This could be achieved thanks to the variable inlet valve diagram (the Japanese car industry already appear to be working on this). Large gains in energy efficiency could be achieved especially on larger cars.

The problem with this kind of progress (as in the past, with turbo-charging technology) is that the same breakthrough can be used to improve energy and environment efficiency (e.g. to ensure a certain power by less energy) or alternatively, to give the driver more power (by an increased energy consumption and thus more pollution). The lessons from the past years have shown that, when "letting free market forces prevail", the latter (more power and more consumption) is the only way really pursued by the car industry*. It is therefore highly questionable that this trend can be "spontaneously" reversed.

The diesel engine for private vehicles has, despite its bad image, a big potential in terms of energy and emission efficiency. The progress achieved in control and command, combined with the design of better combustion chambers and injectors will give diesel engines the ability to meet severe emission regulations.
Tests have also been conducted on the overall adiabatic performance of such engines. A target of 10% improvement within the next ten years seems to be realistic.

With regard to gasoline engines, new geometry such as flat or yoke engines will bring more compact engines to the market, 20% to 30% lighter than conventional units. Furthermore, compact engines will have lower thermal inertia, an important feature for city driving and will allow for a lower air drag profile.

Evolution of car bodies will be of two types, diminution of drag and reduction of weight. The drag co-efficient (Cx) is of importance, even in city driving. Although modern large cars achieve Cx of slightly less than 0.3 such a figure is not so frequent in smaller cars. More compact engines could give special opportunities for the improvement of Cx.

Today prototype cars show the technical feasibility of very lightweight bodies. Due to the limitation on large series production, very lightweight bodies will appear only in limited series such as sport cars or "funny cars", at least in the medium term.

AMT (Automatic Mechanical Transmission) exists on trucks and buses and show through experiments conducted on city bus fleets, an average 10% energy saving compared with manual mechanical transmissions. However the extent of potential energy savings here depends very much on the skill of

---

* For instance, the average car power in the Federal Republic of Germany was 58 kW in 1986 compared with 53 kW in 1981.

the driver. For example for an average private "bad" driver the potential saving may reach 20%. For a very skilful driver the potential saving will approach zero. At present the cost of AMT is still high for private cars: however such a system might be developed for sale on high price cars in the coming years. Here again no-one seems to know whether AMT will be accepted by drivers in preference to the existing system.

In the medium and long term, controls will bring the capacity for further adaptation specifically compensating for the transient period where today's engines are out of tune (e.g. full and mid-power acceleration, driver's mistakes such as accelerator pumping etc.).

C. SOME ISSUES FOR DISCUSSION

Considering the spontaneous evolution of the market, and without a significant shift in the car industry's attitude to energy efficiency, the following consequences may arise:

- a rapid increase of energy demand
- an increase of pollutant and greenhouse gas emissions
- drastic measures at the level of traffic management.

Experience of recent years has demonstrated clearly enough that most technological breakthroughs in vehicle and engine design, once they are developed s a marketable product, result in giving the driver more power (with more energy consumption) rather than the same power with less energy consumption. The case of turbo-charging technology is, in this regard, typical. The problem therefore seems to be at the marketing and commercial level more than at the technological stage.

As far as the automobile in the urban environment is concerned, Italian experiences like Milan and Bologna show that we are already (or are going to be soon) faced with problem of physical compatibility of cars. Moreover, one should bear in mind that any standard on exhaust gas emissions is based on a global level of acceptance (for instance, the ratio between the global quantity of NO_x present throughout Europe and the corresponding air volume should not exceed a certain value). No standard, hard though it is, will ever be able to guarantee that the same level will not be exceeded in urban environments (due to the very high density of cars). Therefore both physical hindrance and environmental concern are key principles in the possibility to admit cars into high density urban areas and the need for selection should be clearly stated. The criteria for this selection will certainly be an important matter for discussion. Whilst one city council could give a priority to resident population, another could definitely opt for road pricing, etc. This latter option already operating in Hong-Kong, Singapore, Bergen (Norway), would also be consistent with the general principle of the "polluter pays".

AUTOMOTIVE FUEL ECONOMY - THE TECHNOLOGICAL POTENTIAL

A Discussion Note prepared by:
Frank Von Hippel
Princeton University, USA

Computer models and the actual construction of prototype automobiles with fuel economies in the range of 2-4 litres/100-km (60-120 mpg) have demonstrated that it is technologically possible to approximately double the average fuel economy of current automobiles without sacrificing interior space and preserving adequate performance. Such large improvements are possible because of the multiplicative effects of:

i) improvements in the efficiency of power plant, and
ii) reductions in power requirements.

With regard to engine efficiency, the principal problem with the currently dominant gasoline-fuelled internal combustion automobile engines is their very low efficiencies at the low fractional power output where most driving occurs. (An automobile engine's peak power is typically only used for very rapid accelerations, for passing manoeuvres at high speed and for pulling trailers up steep hills.

The automotive direct-injection diesel engine has much higher efficiency at low fractional loads. This accounts for most of its approximately 20-45 percent lower fuel consumption on test cycles. The advantage is less at lower vehicle power/weight ratios. Diesel engines have environmental problems (high particulate and nitrogen-oxide emissions) but there continues to be progress in making this engine cleaner. Furthermore, other types of engines which are less developed at this time (e.g. the stratified charge engine) - or "hybrid" systems that would use a less a less powerful engine with peak power being provided from an electric battery or flywheel - could provide a clean, efficient alternative.

Cleaning up the diesel and developing alternative propulsion systems will require significant investments in research and development. In the absence of governmental incentives such R&D is unlikely to be a high priority for the automotive industry. (The low estimates of the potential for fuel-economy improvement reflects the fact that there is currently very little R&D in the automobile industry on changes that will make more than marginal improvements.)

The energy efficiency of the propulsion system can also be improved by ensuring that the engine will be operated at its most efficient speed for each specific power demand at the wheels. This requires a transmission with a wide ratio range within which the ratio can be varied continuously or in relatively modest steps and with electronic controls or a system that indicates to the driver when to change the ratio for fuel economy. Such transmissions have an additional energy-efficiency benefit in that they make available the peak engine power at all road speeds and therefore make possible a reduction in unused engine peak power. In this case, the necessary technologies exist but there is currently no economic incentive to commercialise them.

With regard to reductions of power requirements at the wheels, many different improvements are possible. These include continued reductions in weight through materials substitutions and in aerodynamic drag.

Some of the progress which has already been achieved in reducing power requirements has been offset by a shift towards more powerful engines which operate with poor efficiency in ordinary conditions.

Achieving Large Improvements in Automotive Fuel Economy will Require Public Policy Initiatives

At current fuel prices - and even much higher prices - large opportunities for automotive fuel-economy improvements will remain untapped if the decisions are left to ordinary market forces. The principal reason seems to be that there appears to be a relatively flat plateau in the cost of driving as the fuel economy increases. Efficiency increases will save fuel but, at current fuel prices, the savings are relatively modest as a fraction of the total cost of driving - and they will be largely offset by the cost of the fuel-economy improvements. For example, even if fuel-economy improvements were costless, in a situation where gasoline cost $0.50 litre, only $0.02 km would be saved by halving the fuel consumption from 8 to 4 litres/100 km. For comparison, the fixed costs of owning a car amount to about $0.20/km, assuming an average amount of driving.

Under these circumstances, fuel-economy considerations will tend to be overwhelmed by other attractions in the choice of a new car. Perhaps this accounts for the observed very high rate (20-30 percent per year) average interest rates that consumers are observed to "require" for investments in fuel economy when they typically obtain at best a few percent real rate of return after taxes from investments in average financial instruments.

It therefore appears that, if it is in the social interest to take advantage of the large technological potential for automotive fuel economy improvements, governmental intervention in the market will have to be undertaken.

Increased fuel taxes will have a relatively small effect because of a combination of its relatively weak effect on the flatness of the cost "plateau" and the "irrational" insensitivity of the new car buyer to the economic benefits of fuel economy improvements (20-30 percent required rate of return - see above). Doubling the cost of gasoline in the example above to $1.00/litre, for example, would only double to $0.04/km the incentive to switch from an 8 to a 4 litre/100-km auto.

Therefore, we appear to have a situation in which governmental fuel-economy standards and/or taxes on gasoline "guzzlers" (rebates on "sippers") could result in benefits to both society and the individual automobile purchaser by correcting the irrationality of the current market for fuel economy.

LOW CONSUMPTION - LOW EMISSION PASSENGER CARS

CLAUDE GERRYN

BUSINESS AND INDUSTRY ADVISORY COMMITTEE
TO THE O.E.C.D.

The focus should be on identifying and implementing strategies that have the most potential, yet are both **technologically feasible** and **resource-efficient**. There are important lessons to be learned from the experience of the 1970s, when goals were frequently established without enough attention to **feasibility, timing** or the **impact of conflicting regulations**. In that period, especially in the US, well-intentioned goals became inflexible mandates. These mandates proved to be very costly and led to a deterioration in product function that even prompted some consumers to disconnect emissions systems or delay new car purchases.

° <u>The automobile manufacturers support</u> programmes that are technologically feasible, resource-efficient, and have a positive impact on clean air, global warming, fuel conservation and safety.

  Most of these goals are inter-related to some degree. Some, in fact, directly conflict with one another.

  - While tighter emissions standards can make incremental improvements in air quality, they have a **negative effect** on petroleum conservation.
  - Increased fuel economy standards can **affect safety** and consumer **choice**.
  - Efforts to reduce CFCs and improve safety will **reduce fuel economy**.

Moreover, there is a cumulative impact of regulation that must be considered. If regulatory costs increase too dramatically and rapidly, it would likely encourage consumers **to keep older vehicles** that are less fuel-efficient and do not meet today's stringent auto emissions and safety standards.

The bottom line is that our resources should be focused on the strategies that have the most potential for addressing our goals, providing realistic timetables for new tasks and minimizing the impact of conflicting requirements.

° **New automotive exhaust emission standards** would be resource-efficient provided they:

- Focus on **non-methane hydrocarbons** -- the only ones that affect ozone formation. California already concentrates its controls on these hydrocarbons.

- Recognize factors affecting the emissions performance of vehicles in customer hands that are outside manufacturers' control. For example, **improperly performed maintenance** and failure to perform required maintenance can cause a fleet to **fail** an **in-use** test. The **quality of fuel and lubricants** available to customers varies widely and can reduce control system efficiency.

  Abnormal vehicle operation, such as towing a trailer beyond the weight rating specified in the owner's manual, can overheat the catalyst and accelerate its deterioration. Improper use of leaded fuel and other forms of tampering can seriously reduce the efficiency of the emissions system.

- Provide **sufficient lead time not short-circuited by tax incentives** for new requirements **and a phase-in** of those requirements. Leadtime allows a manufacturer to **refine its designs and processes** before incorporating them into full-scale production, develop the necessary **supply base** and implement new requirements **in step with the normal cycle** of new product introductions, product improvements and engineering **resource availability.** A phase-in period provides time for field testing and allows manufacturers to factor consumer reaction in the development process;

- The combination of numerical standards, effective dates, useful-life and other emissions requirements determines the magnitude of the compliance task for manufacturers.

- It is my understanding that the administration bill -- which would set the car hydrocarbon standard at the **same level as the 1993 California** standard -- is **at or beyond the limits of industry's ability** to comply. Representative Waxman's Bill would set standards twice as stringent as those proposed by the Administration and, at the same time, increase the passenger car useful-life requirement to ten years/100,000 miles.

  Other provisions in each of the bills also pose substantial tasks, for example, proposed truck standards in the Administration and Waxman Bills are beyond industry's projected capability.

COMPARISON OF US (FED) PASSENGER CAR EMISSIONS REQUIREMENTS/PROPOSALS

	Present Req't.	Group of Nine Bill (H.R.99)	Admin. Bill (H.R.3030)	Waxman Bill (H.R.2323)
Hydrocarbon Stds.				
Cert. (gpm)	0.41 THC*	0.25 NMHC**	0.25 NMHC	0.125 THC
In-Use	Same as Cert.	EPA to set	Interim Std.; Phase-in	Same as Cert.
Phase-in Period	-	3 yrs	3 yrs	None
Useful Life (Yrs./Miles)	5/50,000	5/50,000	5/50,000	10/100,000

* Total Hydrocarbons

**Non-methane Hydrocarbons

° Control of Evaporative Emissions and running losses will result in further reductions.

There are four conventional ways to reduce motor vehicle contributions to ambient hydrocarbons beyond the improvements that are assured as new vehicles replace older ones. As noted elsewhere, alternative-fuel vehicles can make further contributions.

By far the largest contribution can be made by controlling fuel evaporation from vehicles on hot days -- so-called running losses and evaporative emissions. Such actions can reduce ambient hydrocarbons, the primary ingredient of smog (ozone), by a total of 25 percent.

- EPA's new rules requiring lower gasoline volatility -- effective June 30, 1989 -- will achieve an 11 percent reduction. Similar requirements are to be expected in Europe as well.

- The remainder of running losses and evaporative emissions can be eliminated by a substantial further reduction in gasoline volatility or through a combination of fuel and emissions systems redesign and lower gasoline volatility.

Controlling vapors during refuelling -- whether with service station controls or with onboard systems (about which there are unresolved safety concerns) could yield only a 1 percent reduction.

Similarly, tighter <u>exhaust emission</u> standards can reduce hydrocarbon emissions by less than 1 percent.

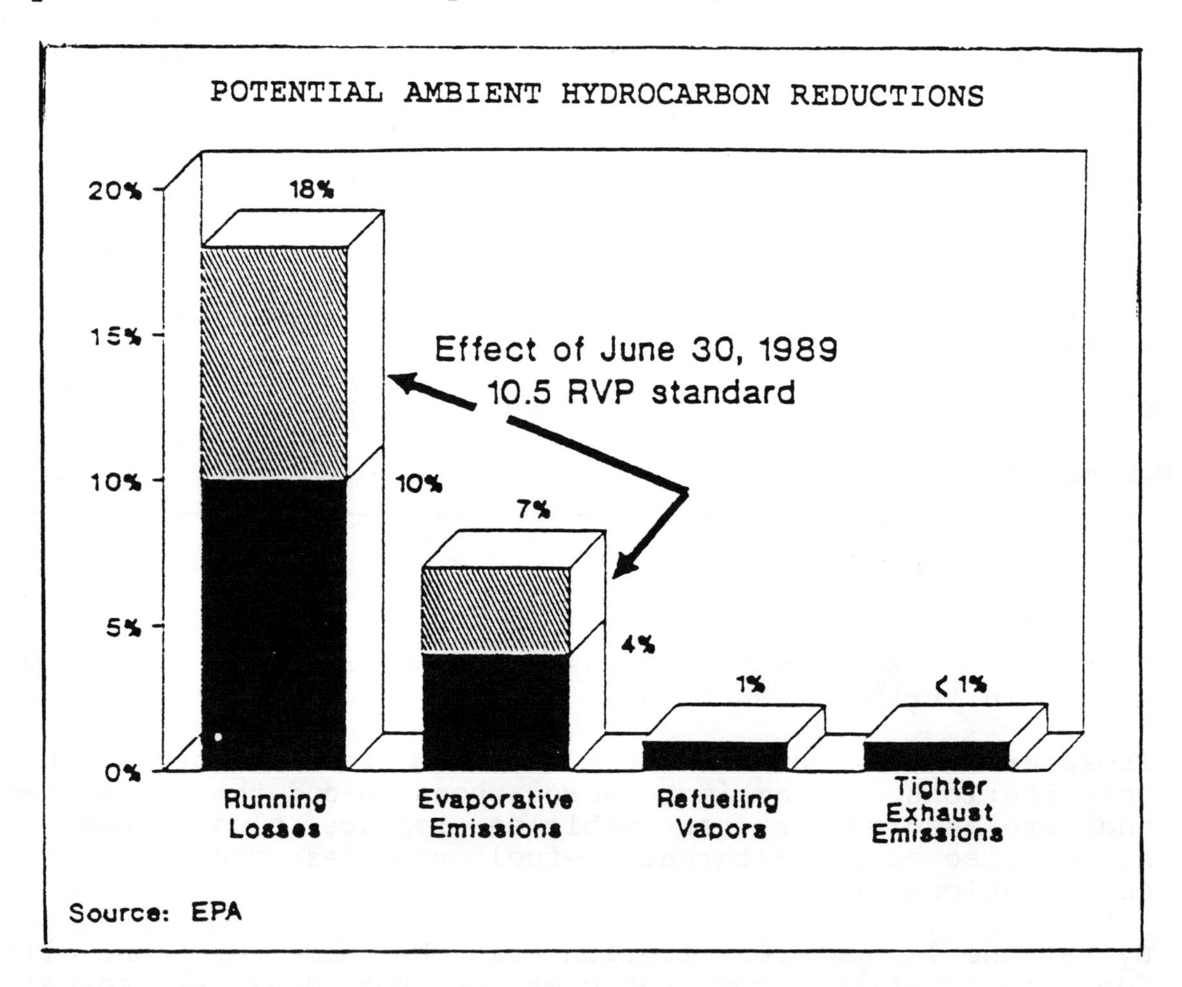

° <u>Reductions in vehicle's Hydrocarbon Exhaust Emission</u> assuming constant Reid Vapor Pressure (RVP)

Substantial and ongoing progress in reducing motor vehicle hydrocarbons is already assured under existing legislation on both continents.

Average motor vehicle hydrocarbon emissions will be reduced by more than 50 percent between now and the year 2000 as new cars and trucks with more advanced emission systems replace older models on the road.

At the same time, vehicle miles travelled (VMT) are projected to increase during this period -- as additional vehicles enter the fleet. This will tend to offset reductions in vehicle hydrocarbon emissions to some extent.

Even assuming that VMT increases by say 2 percent annually, fleet turnover will still account for a reduction of approx. 40% in vehicle hydrocarbons.

- **Workable Alternative Fuels Programs can have combined Clean Air and Energy Benefits**

Alternative fuels -- such as **methanol, ethanol, compressed natural gas** and gasoline that is reformulated to reduce emissions -- can potentially achieve clean air benefits beyond conventional vehicle emissions control strategies.

Methanol fuel seems to be substantially less reactive in the atmosphere and therefore contributes less to smog formation.

A government/industry programme however should be established to resolve technical, marketing and scientific issues before embarking on a high-volume programme. Technical issues include cold weather performance, durability, fuel economics and distribution. Marketing issues involve what it will take to overcome concerns about price, unfamiliar technology and reduced driving range. Scientific issues include air quality benefits and the effects on health and global warming.

WORLDWIDE CONTRIBUTORS TO GLOBAL WARMING

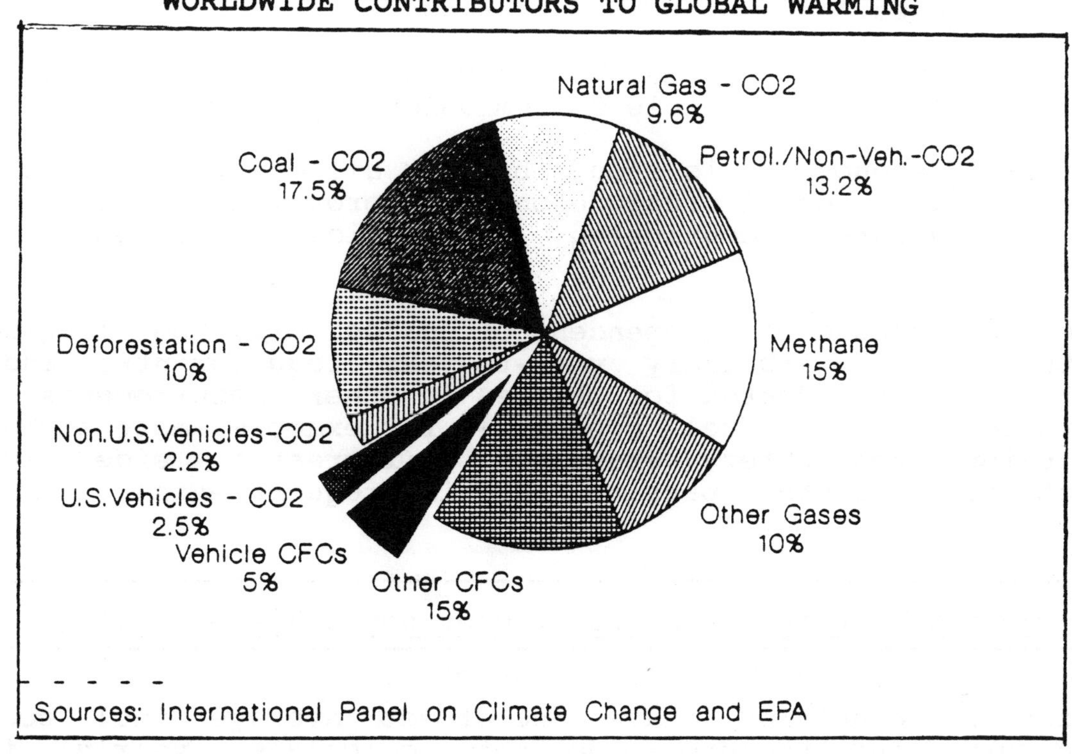

Sources: International Panel on Climate Change and EPA

° **Eliminating CFSs Offers Greatest Potential for Global Warming**

There is general agreement in the scientific community that there is a build-up of "greenhouse" gases in the atmosphere -- primarily chlorofluorocarbons (CFCs) and CO_2 -- that could be contributing to a long-term global warming trend.

Elimination of CFCs -- used in the automotive industry for air conditioning, foam structures and solvents -- is the most significant contribution auto manufacturers can make to reduce greenhouse gases.

It has however to be taken into account that for some uses, suppliers must develop alternatives to CFCs. Testing to confirm the acceptability (e.g. toxicity) of these alternatives has not been completed, and volume production will not begin before 1993.

Re-designing of the air conditioning systems, the largest use of CFCs in US -- to accommodate the most likely CFC substitutes, will be needed. Design modifications will include increasing the size of condensers and evaporators, redesigning compressors, and vehicle changes to accommodate these larger components.

Development of recycling methods to prevent CFCs from escaping into the atmosphere during air conditioner servicing is another task to be successfully undertaken.

In contrast to a CFC elimination strategy, **improving new car fuel economy** would provide only marginal reductions in global warming gases.

In fact, **doubling the fuel economy of the entire U.S. car fleet would reduce vehicle CO_2 emissions by only 0.5 percent.**

If fuel economy goals should be set, all manufacturers should have to face equal fuel economy improvement tasks. This would assure continuing fleet-wide improvement and equitability.

The goals should be independent of sales mix, be long-term, rather than arbitrary year-to-year requirements; and be automatically adjusted for changes in other requirements that adversely affect fuel economy -- for example, more stringent emissions and safety standards. (increased side impact protection and other passive restraint requirements).

Even the European Car Fleet does not achieve 35 MPG

European countries have no fleet average fuel economy standards and therefore do not calculate yearly fleet averages. However, CAFE-type data were compiled by the International Energy Agency in 1985 to measure industry progress toward a voluntary 10 percent improvement goal that was agreed at the time of the oil embargo.

The data show that the fuel economy of new cars sold in Europe in 1985 averaged in the low 30 mpgs -- not substantially higher than the United States, despite major differences in the size and the type of cars.

- About 80 percent of new cars sold in Europe are compact-size or smaller, compared with about 50 percent in the United States.
- More than 90 percent of European cars are equipped with manual transmissions versus 28 percent in the United States.
- Sixteen percent of European cars are equipped with diesel engines, compared with less than 1 percent of U.S. cars.

Moreover, European fleet averages are supported by an infrastructure that includes **high gasoline prices** (about $2.75/gallon), **tax incentives** to encourage the purchase of small vehicles and some extensive, convenient and low-cost **public transit systems**.

What is needed is to continue to introduce advanced fuel-economy technologies. The most promising are multi-valve, overhead cam engines; continuously variable and electronically controlled automatic transmissions; and engines incorporating variable-valve timing.

These technologies should yield important incremental fuel economy improvements. However, there are no **imminent breakthrough** technologies that can produce **major** fuel economy increases. Studies in the US supporting much higher "CAFE" levels often cite concept vehicles that use diesel or two-stroke engines, weigh 900/1,900 pounds, carry two to four passengers and generally fail to meet U.S. safety and/or emissions requirements.

GENERAL INDUSTRY CONCLUSIONS

Where the protection of the environment is an agreed objective of business and industry, identification of clear and stable objectives, technologically feasible and resource-efficient with much attention to flexibility, timing and impact of conflicting regulations must be seen as the key ingredients for success.

Most of the projected goals which are expected to have a positive impact on clean air, global warming, fuel conservation, noise and safety, are inter-related to some degree, sometimes in fact directly conflicting with one another.

This clearly underlines the need for a global approach, bringing the problem-solving factors as well as the burdening factors into the equation.

While tighter emissions standards can make incremental improvements in air quality, they have a negative effect on fuel conservation. Efforts to reduce CFCs and noise and to improve safety will reduce fuel economy. Moreover, there is a cumulative impact of regulation that must be considered. Increased fuel economy' standards can affect safety and consumer choice.

If regulatory costs increase too dramatically and rapidly, it would likely encourage consumers to keep older vehicles which are less fuel-efficient and which do not meet today's stringent auto emissions and safety standards.

The bottom line is that our resources should be focused on the strategies with the most potential for addressing our goals, providing realistic timetable for realistic new tasks and minimizing the impact of conflicting requirements.

New automotive exhaust emission standards would be resource-efficient provided they :

- Focus on non-methane hydrocarbons, being the only one which affect ozone formation.

- Recognize factors affecting the emissions performance of vehicles in customer hands that are outside manufacturers' control; improperly performed maintenance and failure to perform required maintenance.

The quality of fuel and lubricants available to customers varies widely and can reduce control system efficiency.

Abnormal vehicle operation, towing a trailer beyond the specified weight rating, can overheat the catalyst and accelerate its deterioration.

Improper use of leaded fuel and other forms of tampering can seriously reduce the efficiency of the emissions system.

As for any time-consuming engineering exercise, sufficient lead-time not short-circuited by tax incentives for new requirements and a phase-in of those requirements are imperative. Lead-time allows a manufacturer to refine its designs and processes before incorporating them into full-scale production, develop the necessary supply base and implement new requirements in step with the normal cycle of new product introductions, product improvements and engineering resource availability. A phase-in period provides time for field testing and allows manufacturers to factor consumer reaction in the development process.

The combination of numerical standards' effective dates, useful-life and other emissions as well as non-emissions / fuel economy requirements determines the magnitude of the compliance task for manufacturers.

What is needed to lower fuel consumption is to continue to introduce advanced fuel-economy technologies. The most promising are high power density engines (multi-valve engines, sophisticated "engine breathing" control systems), electronically controlled continuously variable transmissions.

However there are no imminent breakthrough technologies which can produce major fuel economy increases. As a matter of fact, the adoption in Europe of emission limits equivalent to those in the US has lead to a diminishing interest for alternative technologies such as the lean-burn engine concept.

The diesel, which has become a "European Technology", is at present seen as the most fuel efficient engine available. While further improvements, through electronic control, could enable the industry to meet the presently proposed European standards in 1992, they would have to be totally abandoned if tighter standards on nitrogen oxide are adopted.

Some further progress will be achieved by reduction of weight, air and rolling resistance through progress in design methodologies and scientific advance in electronic and material technologies.

Improvement of the traffic system and road network, combined with a stronger awareness of the negative effects of some owners' / drivers' behaviour through information and education will greatly help to resolve this global "low pollution - low consumption" task.

Even the much-praised need for transition towards smaller cars, lower engine displacement and reduction in the engine power output, have to be agreed by our customers first.

After all, a continuing, open dialogue between governments and the public sector, the petroleum and automobile industries is fundamental to any valid problem-solving definition. Not short-term reversible policy decisions, motivated by electoral demagogism, but long-term orientations with specific targets, and including economic impact / aspects, will bring about realistic and feasible solutions.

Clean and Low Consumption Automobiles

J. van der Weide, R.C. Rijkeboer, P. van Sloten

TNO Road-Vehicles Research Institute
Delft, the Netherlands

1. Introduction

One or more decades from now it may be expected that small electric vehicles will be the most usual means of transport in city traffic. Not unlikely is that for intercity driving hydrogen fuelled vehicles with automatic pilot systems will appear on highways. Even a kind of coupling of these vehicles in order to reduce air resistance may be common at a certain moment. It will be imperative that for this kind of transportation system an immense change in mentality as well as in infrastructure will be required.

This presentation, however, will deal mainly with selected topics which should be part of the R&D effort in the coming decade. Although the need for a complete change of attitude in physical planning in order to reduce the need for transport is understood, the paper will be limited to the field of experience of the authors. This being the energy and emission aspects of the internal combustion engine and its various fuels. Discussed are the topics: Laws, Directives etc., Automotive Technology and Fuel Technology.

2. Laws, directives, etc.

Considering the fact that automobile manufacturers concentrate on building engines and vehicles which will sell on the market place, there will be a need for laws and guidelines which will direct the automobile manufacturers and the public towards the early introduction of cars with a lower fuel consumption and lower emissions and subsequently the use of these vehicles. The testing of about 450 vehicles-in-use in the Netherlands confirms the statement. Before tuning only one third of the vehicle types complied with their respective standards. Even after tuning about 15% of the types still failed to meet the standards. The development of experimental low consumption/low emission vehicles by independent institutions is required to set the pace for more severe regulations in the near future.

Present regulations are covering only partly the climatic circumstances in which vehicles are actually used. Vehicles tested, for instance under low temperatures (0°C), show very often much higher emissions. Changing the prescribed test temperature from 20-30 °C to for instance 0-30 °C should lead to more R&D on emissions at low temperatures and in the end to lower overall emissions.

In order to enable the calculation of national or global pollution emission factors were established for each kinds of source. From the testing of vehicles-in-use it is apparent that emission factors for automobiles need to be updated regularly due to changes in technology. In particular emission factors for heavy duty vehicles are almost non-existent. To say the least the ones in use are completely unreliable. As a basis for national and international pollution abatement policies these emission factors should be reliable, up-to-date and should reflect the emissions of vehicles in actual use. It goes without saying that more effort should be put in this work.

In a more detailed form than used for global or national pollution calculations these emission factors could be used to reduce emissions in local situations such as city centres. Both as a design roll for a local infrastructure with minimum emissions and in computerized traffic control systems which control the actual traffic flow. As far as known R&D in this field is minimal and should be increased shortly.

In most studies where a comparison is made of the emissions of vehicles driven by different fuels or comparisons of different modes of transport, the comparison does not include the energy consumption and emissions from the production of the fuels and the vehicles. Also the transport of these fuels and vehicles to their place of use is excluded. These contributions to the global pollution should also be taken into account by policy makers when discussing the merits of shifts in conventional fuel use and before the introduction of possible alternative fuels, such as methanol, ethanol, natural gas, reformulated gasoline, etc. A comprehensive survey of possibilities, including energy consumption and emission figures for each step in the chain from production to final use will be of great value in particular for policy makers in the transport and physical planning sector.

3. Automotive Technology

With regard to technological developments on vehicles themselves the following observations can be made. The spark ignition (Otto cycle) engine is the main propulsion unit for passenger cars, and that for very good reasons. Its known disadvantage relative to the compression ignition (Diesel cycle) engine is its higher fuel consumption, resulting also in a higher emission of CO_2. When this is analyzed further it turns out the maximum efficiency is not at all that different, but that it is the drop in efficiency at part load which is more pronounced for the Otto cycle than for the Diesel cycle. A much improved load efficiency for Otto cycle engines is therefore a priority item. This may well lead to an integration of the Otto cycle and the Diesel cycle, producing e.a. a spark ignited qualitativelty controlled cycle (without the efficiency destroying throttle valve) that combines the virtues of the two principles. Such combustion systems have been proposed in the past for reasons of multi-fuel capability, but could be pursued in the future for higher efficiency purposes, even if this should mean the use of a dedicated fuel.

Especially on an engine with qualitative control (mixture strength control), rather than quantitative control (charge volume control), variable compression ratio would be of great benefit. It would raise part load efficiency still further. This could be combined with improved control over the mixture charge through variable valve timing etc. Another approach could be to switch off cylinders at part load, so that the remaining working cylinders work nearer full load. This approach had a short period of popularity after the oil crisis. At that time it was performed by opening the valves of the non-operating cylinder.d(s), so that the piston just pumped gas, without firing. A more advanced approach would be to stop one or some of the pistons altogether, thereby eliminating all or most of the mechanical losses as well.

Further use of available fuel energy could be made if it was attempted to extract more energy from the expanding gases, e.g. by compounding. Especially when cylinder switch-off is used, the near full load condition of the remaining cylinders might make this a viable. Minimisation of heat losses from the cylinder would further enhance it. Minimisation of heat loss would also be beneficial for the quick warming up of the engine, thereby cutting the excess emission and fuel consumption of the cold start phase.

As far as the vehicle as a whole is concerned a minimisation of propulsion energy needed would further reduce consumption and emission. Especially in fluctuating traffic conditions, such as in urban areas, kinetic energy forms a large part of the propulsion energy required. When vehicles are fitted with a form of energy storage device regenerative braking becomes a possibility. Under such circumstances this can lead to substantial reductions in consumption and hence emission. More advanced forms of energy storage would allow the engine to run at a near constant rate while the energy storage system would deal with the fluctuations in the propulsion energy required. Such a hybrid system would partially be an alternative for some of the engine developments indicated earlier. If the energy storage capacity is large enough this would allow e.g. electric propulsion in urban areas which could alleviate local pollution problems.

4. Fuel Technology

In the Netherlands experience is gained since the fifties with LPG as an automotive fuel. Now some 700,000 vehicles are running refuelled from some 2,000 filling stations. More recently natural gas is put in use in mainly converted diesel engines. This makes it possible to meet US 1991/1994 emission standards for heavy duty engines. In the longer term it is foreseeable that the use of in particular natural gas will increase due to:

- low emission
- lower CO_2 emission
- good availability.

Due to the need for even more emission reduction and CO_2 reduction the use of non-hydrocarbon fuels can be expected. In theory this can be hydrogen, ammonia or peroxides.

The experiences with alternative fuels so far has learned that its introduction is very difficult. This is due to the fact that car manufacturers do not produce cars when there is no reasonable infrastructure for a new fuel and on the other hand an infrastructure for new fuels is not built when it is not clear that cars will be produced. This is a kind of chicken and egg dilemma. With the introduction of LPG in the Netherlands this problem was and is overcome by keeping the vehicles switchable to petrol in case the vehicles has to operate in an area where the LPG is not available. The application of natural gas in city buses for instance is possible without the possibility to switch to diesel operation due to the fact that such buses are fleet operated and turn back on their base once or twice a day. Demonstration projects with methanol in America and West-Germany clearly demonstrated that running dedicated methanol vehicles with a poor infrastructure of filling stations was very problematic. Next demonstrations with methanol in America will be with so called FFV's (Flexible Fuel Vehicles). This means that such vehicles can be refuelled (in the same tank) with petrol when necessary. The foregoing leads to the conclusion that for large scale introduction of alternative fuels a kind of flexible fuel operation is necessary, at least for a gradual introduction of new fuels.

It is further thinkable that different kinds of fuels can be handled in one special tank in the vehicle. This may include gaseous fuels as well as liquid fuels. Such mono tank systems have the advantage of weight and volume reduction. This leads to the conclusion that a kind of universal tank has to be developed.

Some alternative fuels, in particular hydrogen, are difficult to store in sufficient quantity on board of the vehicle, large quantities of some alternative fuels on board could also be a safety hazard. Therefore it would be attractive to store a small amount. This than has to be compensated by a very fast and simple refuelling. This could be done by an automatic system which refuels the car in some 20 seconds without a person leaving the car and with automatic payment via electronic vehicle recognition. This leads to the conclusion that a kind of robotic refuelling with different fuels in one universal tank has to be developed.

5. Summary of recommendations:

5.1 Laws, directives, etc,

* Development of experimental low consumption/low emission vehicles
* Reduction of the ambient test temperature to e.g. 0 °C
* More and better emission factors for different modes of transport
* Better survey of energy consumption and emissions of all the links in the chain from source to end-user.

5.2 Automotive Technology

* R&D on the integration of Otto and Diesel cycle
* R&D on variable compression ratio's, possibly in combination with switching-off cylinders and variable valve timing
* R&D on compounding and reduction of heat losses in combination with the foregoing topics
* R&D on recuperation and storage of braking energy and the development of hybrid systems.

5.3 Fuel Technology

* R&D on natural gas engines for vehicles
* Development of engine management systems which can handle more than one fuel (Flexible Fuel Vehicle)
* Development of a universal on board tank which can contain different fuels, even in a mix
* Development of robotic refuelling with automatic payment.

THE POTENTIAL FOR IMPROVING FUEL CONSUMPTION

KARL-ERIK EGEBÄCK

THE SWEDISH MOTOR VEHICLE INSPECTION COMPANY

INTRODUCTION

Concerning the energy consumed by automobiles and hence, pollution emitted, there are at least four areas which can easily be defined and within which measures could be taken to improve the current situation.

Progress could be achieved by;

* improvement of the vehicle
* improvement of the traffic system and road network within cities and countries
* information to and education of the car owner or driver
* use of fuels which form less pollution when combusted and which are more energy efficient and hence emit less CO_2 and/or other greenhouse gases

In this paper some of the measures which can be taken to reduce the automotive emissions will be discussed briefly.

IMPROVEMENT OF THE VEHICLE

When looking at the improvement of the design of the vehicle the following measures can be of greatest interest;

* a better aerodynamic design
* tires with low rolling resistance
* a lower curb weight of the vehicle
* smaller vehicles (passenger cars)
* higher engine efficiency
* restricted idling of the engine
* engines with less deplacement
* use of a transmission which is matched to the engine
* improved automatic transmission
* use of a six gear gearbox to reduce the engine speed

*	use of electronics to guide the driver when the traffic is congested

*	improved cold start behaviour by use of heaters for the engine and the catalyst

Most of these improvements can be achieved within 5 to 7 years and some of them within 3 to 4 years. In the manufacturers laboratories extensive work is going on to improve the engine efficiency, the transmission, the aerodynamic design, the rolling resistance etc. Unfortunately there seems to be a strong demand among those who purchase passenger cars for more powerful and larger vehicles. Therefore there is no clear incitement for the car manufacturers to manufacture smaller vehicles with less powerful engines. The attitude among the car owners has to be changed so as to be more oriented towards smaller and less powerful vehicles. Despite the fact that there may be a conflict between the size of the vehicle and the safety for the driver there must be a potential to reduce the size and the weight of the vehicle.

The average number of passengers (including the driver) in a car is less than two. Also for larger vehicles the fuel consumption can be reduced to a great extent by redesigning the transmission so as to keep the vehicle running at the same velocity at a lower engine speed. An increase of the gear ratio in order to increase the speed of the vehicle from 40 to 60 km/h at an engine speed of 1000 rpm may decrease the fuel consumption by approximately 25 %. The disadvantage of such a measure is that the engine may not run smoothly and the car owner will then complain.

The different measures which can be taken to improve the fuel economy of the vehicle will have different effects. Based on available literature and experiences gained the potential for reduction of fuel consumption for passenger cars has been estimated to be;

-	up to 20% for higher engine energy efficiency
-	approximately 10 % for improved aerodynamic design
-	approximately 15 % for optimization of the transmission
-	approximately 10 % for reduced weight and size of the vehicle
-	up to 10 % for reduced rolling resistance

Because there is an interaction between the different measures which can be taken the percentages of reduction cannot be added. Taking all of the measures listed above into account the fuel consumption can be reduced by 30 % within 4 years and up to 50 % within 7 years. The use of new vehicle drive systems such as hybrid systems will effect the fuel consumption even more. It is doubtful whether such systems will be in use within a time-limit of 7 years.

It should be pointed out that there are limitations in the above listed measures which are related to driveability and safety of the vehicle. Such a drawback as poorer safety cannot be neglected. On the other hand an investigation carried out by a Swedish insurance company has shown that the optimum vehicle weight with respect to safety is a curb weight between 1000 and 1100 kilograms. Concerning the driveability the car owner may have to accept a poorer driveability in order to get a car with an extremely low fuel consumption.

In Sweden low emission passenger cars have been introduced as a result of emission requirements. The experience of these cars is that the fuel consumption has been improved which is assumed to be a result of the introduction of engines with more efficient fuel systems. It should also be pointed out that in Sweden there is a requirement for the car manufacturers to present data of fuel consumption for each model of passenger cars introduced on the market. Also the customer of the car has to be informed.

Concerning heavy duty vehicles roughly the same measures can be applied in order to reduce fuel consumption similarly to passenger cars. On a short time scale the potential for reduction of fuel consumption is lower for heavy duty vehicles than for passenger cars. On the other hand planning of transportation of goods etc may have a great impact on the energy used and the emission for heavy duty vehicles.

Fortunately in the case of heavy duty vehicles the fuel efficiency of the vehicle is one of the primarily factors for the manufacturer to be used in his competition to attract buyers of the vehicle. In the near future these means also will apply to the emission efficiency of the vehicle. In Sweden emission requirements for heavy duty vehicles will be introduced voluntarily from the 1992 year model and mandatory from the 1994 year model.

IMPROVEMENT OF THE TRAFFIC SYSTEM AND THE ROAD NETWORK

Despite the fact that the work of the OECD/IEA informal expert panel primarily seems to be oriented towards vehicles, a few words should be said about other measures viewed in the introductory remarks. Evaluations carried out in Sweden have shown that extensive improvements of both emissions and fuel consumption can be achieved by improving the traffic system and the road network. Unfortunately to improve the road network is a costly and not always possible measure to be taken.

A well functioning public transportation system with low emission buses in combination with low emission service vehicles (i e vehicles carrying goods etc) in city areas will not only improve the air quality but also make the public transportation system more attractive from an emission point of view. Measures taken to improve the public transportation system and to limit the number of private cars in city areas will have an effect on the traffic situation. The risk of traffic congestion will be less and hence the vehicles can be run smoothly at a constant speed. In addition to this type of traffic planning the introduction of a system for traffic information to the driver of the vehicle also will be a possibility for reducing traffic congestions.

THE CAR DRIVER

The car driver should be informed and educated to drive the vehicle smoothly and especially to avoid violent accelerations in order to save fuel. It is an important task to find ways to make the car owner more aware of the fact that he can, to a great extent, contribute to the effort to improve the situation concerning energy consumption and air pollution.

ENERGY EFFICIENT FUEL

One way to improve the efficiency of the engine is to increase the compression ratio. Such an engine may need a more knock resistant fuel. Use of pure alcohol or alcohol in gasoline will increase the octane number. As the benefit of using alcohol as an automotive fuel is well known, it need not be further discussed in this paper. The fact that use of alcohol will increase the emission of aldehydes and alkyl nitrites must be taken care of by use of efficient catalysts.

POLITICAL/REGULATORY MEASURES ETC

Many different measures can be taken and may have to be taken to improve the situation concerning the energy consumption and air pollution caused by automobiles. Taxation is one measure widely used. Tolls to be payed before entering the border of the city and high parking fees are other ways to restrict the number of vehicles in the city area. All of these measures are apprehended as unfriendly by the car owners and some of them cannot be used to a great extent for political reasons. To pay a very high tax for the fuel may be too heavy a burden for people who, for example, live in the north of Sweden where they have to use their own cars because the public transportation system is not available everywhere.

In that situation other ways have to be found. One way is to work out agreements between the authorities (or the government) and the car industry to step up the introduction of more fuel and emission efficient cars. A threat of penalty may have to be one part of the agreement. Another way which has successfully been used in Sweden is tax incentives as a means of encouraging the introduction of more efficient vehicles.

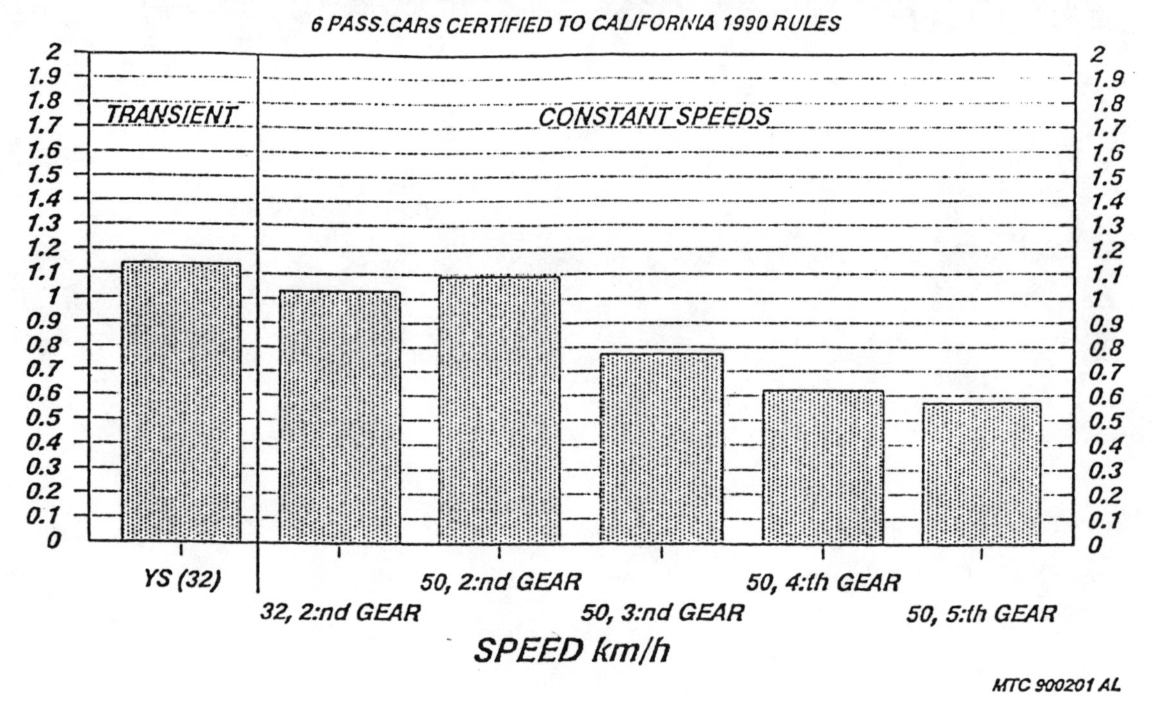

Figure 1

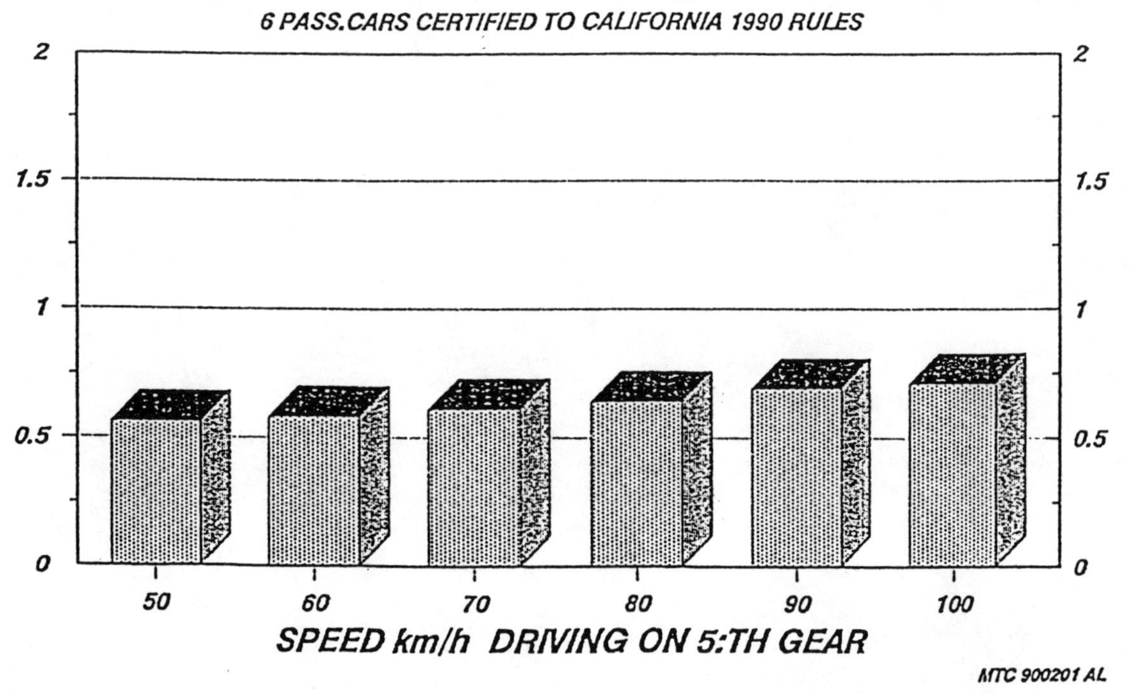

Figure 2

Note to the OECD/IEA Informal Expert Panel on Low Consumption/Low Emission Automobile

Antti Saarialho

Helsinki University of Technology, Finland

1. BACKGROUND

Two of the three aspects of "Background information", in the IEA/OECD letter attachment (20.Nov.1989) calling experts to the Rome meeting, require some comments from the point of view of Finland, at least:

1) According to our statistics the specific fuel consumption of motor vehicles has not remained constant in the last decade:

Finland	1980	1987
Gasoline sales	1 811 000 m3	1 698 000 m3
Motor cars	1 225 931	1 698 671
Gasoline use per car	1, 48 m3/year	1,38 m3/year

2) The yearly mileage per car has also not increased in Finland. In 1980 an average car was driven 19 800 km/year and in 1987 17 300 km/year. The yearly mileages of trucks, buses and pick-up vans have remained virtually constant during the same period of time.

3) The total Finnish consumption of oil products decreased in 1980 to 1988 from 1 343 000 toe to 9 450 000 toe, ie. 16,7%. During the same period the transportation's share of oil energyuse increased from 2 609 000 tons (23 % of oil energy) to 3 576 000 tons (37% of oil energy), the increase 966 000 tons being as well 37 %. At the same time the total amount of motor vehicles increased from 1,390 million units to 2,034 million, ie. 46,3 %, the increase of passenger cars being 47 %.

2. SOME CONSIDERATIONS ON THE DEVELOPMENT OF THE MOTOR CAR

The development of motor cars is today done under a great pressure of the surrounding world and with conflicting goals. Any new car model must be a balanced product in respect to, at least, eight areas of development: energy efficiency, emissions, safety, comfort, reliability, noise level, performance and costs of manufacture and usage.
The development takes time, although the major car manufacturers each have at their R&D-centers some 5000 - 6000 full-time experts. The majority of them have academic degrees and eg. at VW nearly 800 of them are doctors of technology or philosophy!

Even for those experts the development of a new model takes 4-5 years of hard work because they have to take into consideration a great number of variables. Many of the variables are well beyond the control of any technician!
2 (3)

To condense the situation as short as possible the following "equation" can be written:

FUTURE DEVELOPMENT OF THE AUTOMOTIVE TECHNOLOGY

"FDAT" = f(3C + L + 4E + 2M)

Where the **3 C:s are : CONSUMERS, COLLABORATORS, COMPETITORS**

L is: The LEGISLATOR

4 E:s are : EMISSIONS, ENERGY, ECONOMICS and the development of ELECTRONICS

2 M:s are : development of MATERIALS SCIENCE and MANUFACTURING TECHNOLOGY.

This means that the future development of our cars is primarily dependant of the efforts of the R&D-personnel of the major automobile manufacturers - not so much of the efforts of anybody else.

When doing this R&D the the personnel has to take a very close look into the customer demands, capabilities of the component manufacturers and the movements of the competitors. In their work the R&D-personnel further has to anticipate legislational changes especially in respect to emissions and economy. Special attention has to be devoted to the global availability and nature of energy for the transportation purposes. Further they have to take advantage of the rapid development of electronics, materials technology and manufacturing methods.

3. ATTENTION TO THE TIME PERSPECTIVE

When considering the possible speed of any major change in the transportation technology we should bear in mind that in the past it took a very long time to change the propulsion :

- in the ships some 40 years
- on the railway some 30 years
- in the aviation about 20 years ,

for the road transportation it will take, even with the present level of technology, at least 10 to 15 years before any major alternative to those forms of propulsion we know now - or can anticipate with the present knowlegde - could penetrate. Further it should be noted that any alternative has to be competitive on its own merits on markets, where also the earlier solutions, ie. the gasoline or diesel engine, are also being rapidly developed. This is the case even with the use of modern aids for the R&D within the industry!

4. RECOMMENDATIONS FOR FUTURE CO-OPERATION

The expertise within the automotive industry, the fuel industry and research institutes in the OECD countries is immense. All efforts should be made to bring the experts of these fields together to elaborate the topic of the proposed meeting. The author feels especially important that the experts of the energy producers and the automotive industry would take part to the meetings from the beginning, - not just "sometimes later". The question is how they could be tempted to do so!

The OECD should take all the measures to bring together the best brains of the nations to create a program for development. It is hard to believe that any real answers for the questions of the problems of the transportation engineering could be found without of the engineers and researchers of this special field of engineering!

Transportation of goods and people is causing a certain load to the nature. This is admitted even by any engineer working for this sector of technology. To get the engineers committed to the work for a better environment is not difficult, but, according to the author's opinion, it would require sound scientific proof for many of the claims expressed towards the use of motor vehicles today. The transportation is causing air pollution, there is no doubt of that, but how could we get along without transporting goods and people? How much better would the air quality be in our countries if we refrained from the use of the motor car ? The author firmly believes that it is high time to put the air pollution question, inclusive that of the CO_2-emission, into scientifically solid perspective. The possibly harmful effects of the automotive technology can be remedied - but only with the help of better technology!

5. RECOMMENDATIONS FOR ACTIONS

The following is a short list of desirable lines of research for all partners concerned:

1. **Development of the electric propulsion for road vehicles**
2. **Development of the hydrogen technology for road vehicles**
3. **Promotion of public transportation in densely populated areas**
4. **Promotion of the use of LPG and LNG in city traffic.**

The two last mentioned efforts would have the quickiest effect on the air quality in congested town areas, where also the greatest number of people is subjected to the health effects of the air pollution from motor vehicles. At the same time the efforts of automobile manufacturers in building new and better models should be appreciated and promoted.

According to the opinion of the author the OECD expert panel should concentrate its efforts on the above four fields of R&D and bring forward good co-operation between all interested partners.

The fuel-efficient lightweight car

Centre for energy conservation and environmental technology, Delft, The Netherlands, 31-15150150

A.N. Bleijenberg, B.J.C.M. Rutten

Historical trends

In the past ten years cars became more efficient. An insight can be obtained based on the ECE figures. Figure 1 shows the fuel-consumption as a function of empty vehicle weight; by means of regression-techniques a lineair correlation was derived.

State-of-the-art 1989; based on ECE15, 90, 120 km/h

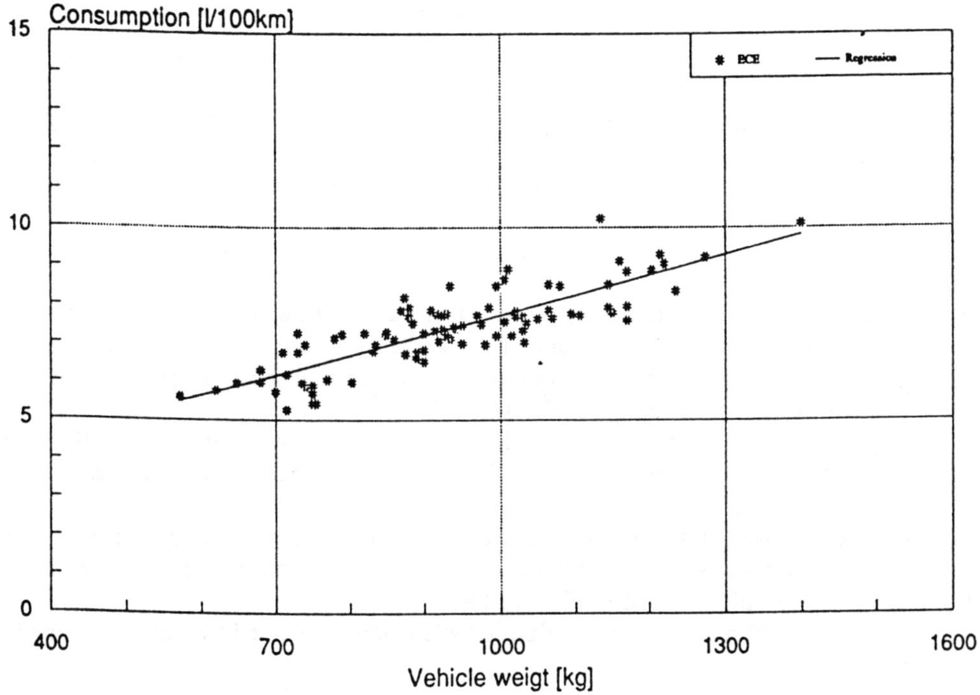

Fig.1 Fuel-consumption gasoline cars; state-of-the-art 1989

This was not only done for the state-of-the-art of 1989 but for 1980 and 1984 as well; see figure 2. In the period from 1980 to 1984 light cars in particular became more efficient, about 11%; heavier cars some 7%. In the period from 1984 to 1989 smaller cars have hardly increased their efficiency, only some 3%, whereas heavier cars did again improved by around 6%. The average efficiency improvement is about 14%. In figure 1 there are also some high fuel economy prototype vehicles shown. The low fuel consumption of these cars is mainly a result of the the very low C_w-factor of 0.20 and the low vehicle weight.

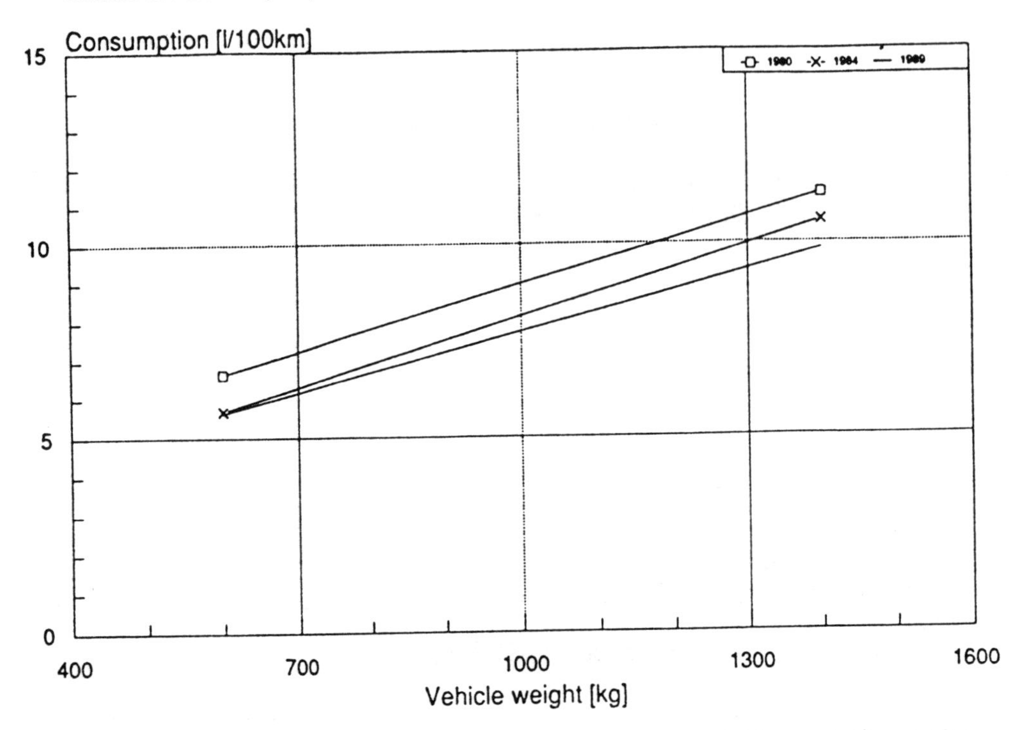

Fig.2 Fuel-consumption gasoline cars; state-of-the-art 1980, 1984, 1989

The total fleet of cars in The Netherlands however, has become less more efficient over the past ten years.
One reason for this is the increase of the average weight from 900 to 920 kg (2%), resulting in 1.5% higher fuel-consumption.
A second reason could be the different source material, technical and statistical.
A third reason is the strong increase in engine power.
Figure 3 shows the function of specific enginepower (enginepower devided by empty vehicle weight) and vehicle weight. A remarkable increase of 25% for small cars and 15% for big cars can be detected.
Also, an estimate has been made of the specific power of the ECE test-runs at a constant speed of 90, 100 and 120 km/h and at acceleration from 0 to 50 km/h in the ECE-15 cycle.
From figure 3 the conclusion can be drawn that the power demand in the ECE-cycle is hardly representative as compared to real driving practice (more representative will be the US-FTP cycle).
One keeps in mind, that the specific enginepower of city and intercity busses is about 12 to 15 W/kg.

In general more engine-power gives one the possibility to use a car in a more fuel-consuming driving-style:

- especially when accelerating fast and slowing down firmly at the last moment instead of decelerating by resistance of the air and rolling out the vehicle with the engine idling. The result of this driving style is a higher average speed, resulting in a higher fuel consumption;
- one will be invited more easily to drive with higher speeds, comparing with a car with less engine power.

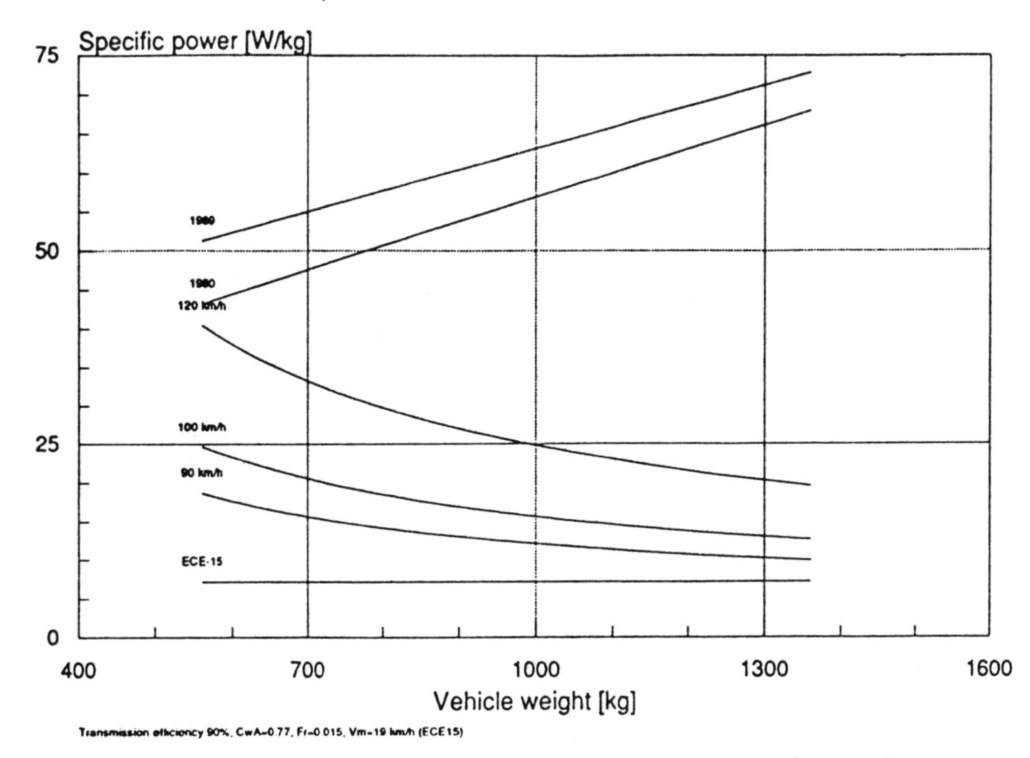

Fig.3 Specific enginepower gasoline cars; state-of-the-art 1980, 1989

Technical developments.

The main research and development effort is focused on getting the highest efficiency η in all circumstances. We call lower C_w-factors, engine-management systems, higher efficiency of engine and transmission components, electronically controlled CVTs etc. The most important technical developments in the next five years will be:

- through the use of plastics and lighter materials the weight of cars will be reduced by some 5%, resulting in a 3% decrease in fuel-consumption;
- the maximum engine-efficiency may increase by several percent, far more important however, is the operational efficiency. We regard a maximum increase in operational efficiency of 1.0 to 1.5% to be realistic, using multiple valve techniques and turbochargers, resulting in a higher efficiency at partial loads. This will result in a decrease in fuel-consumption of 5 to 10% (average efficiency nowadays is about 15-20%);
- a more than marginal reduction of resistance of the C_w-factor in the period mentioned seems unrealistic because of long development times and high costs; the average C_w-factor nowadays for new cars is 0.33 (0.28-0.38) and can be lowered to 0.30 (0.27-0.33);
- use of a CVT has a potential for reduction of fuel-consumption in town-traffic, not on through-roads. Only if the CVT is tuned in such a way the engine will operate at its highest efficiency, this will result in a more than marginal reduction of fuel-consumption. This means the maximum acceleration power will not always be available!

For the year 2000 we can add:

- a further reduction in weight of again 5%;
- possibly a higher operational efficiency of the internal combustion engine.
- in the whole fleet a more then marginal decrease of the C_w-factor to about 0.30.

Only after 2000 there will be possibilities for the use of a regenerative brake-system by means of fly-wheel and CVT-technology. It is unlikely these will be commercially available sooner. Savings of 10 to 20% are then feasable; The C_w-factor will probably decrease to 0.20-0.25.

Fuels

In the next five years more use of LPG could be reduce the CO_2-emissions by 10%-15%, compared to a gasoline engine. Using methane, this reduction will be 25%-30%. In this case important issues to be solved are:

- availability of light fuel tanks;
- use of dual fuel engines (gasoline and methane)
- good infrastructure of methane pump stations

Later still, between 2000 and 2010, durable fuels may become available, especially important for the decrease of CO_2-emmisions. A non-CO_2 energy conversion will result from the production of hydrogen by means of electrolysis by durable energy sources like wind-, solar- and hydro-power. By way of the methanol-route, produced by especially planted bio-mass, a CO2-cycle will result.
From a CO_2 point of view the electric car would only be worth considering if the necessary electric power can be drawn directly from the grid or would be produced by durable energy sources. (The efficiency of the electric energy taken from the grid will grown to about 45%-50% -e.g. by use of Combined-Heat-Power plants- in the future. The total energetic efficiency of the vehicle with internal combustion engine nowadays is about 15 to 20%. In the future after the year 2000 figures of 20 to 30% can be reached). In a hybrid system the use of fossil fuels will remain the same or may even increase as a result of extra energy-conversions.

Summary of technical developments

In summary, in the short term until 1995 the average car may become some 10% to 15% more efficient, about 15% to 20% by 2000. The key to the realisation of the target of 30 to 50% reduction of fuel-consumption lies in a totally new concept of cars:
A light-weight car (400 - 500 kg) with low specific power and sufficient acceleration power; 25 W/kg (90 km/h) to 35 W/kg (100 km/h), in other words 10 kW to 20 kW, depending on the vehicle weight and the maximum speed desired. Together with the use of light-weight materials, low Cw-A factors and modern engine-management systems this may well be the high-tech car of the future with a fuel-consumption (ECE) of 3 to 5 l/100km.

Fuel-consumption total car fleet in The Netherlands

In the past three years the amount of vehicle kilometers totally driven in The Netherlands rised with 14%. In the National Environmental Policy Plan the national government made prognostications for the next 20 years, that the average growth per year will be not more than 1.7%, and that the total CO_2-emisions will be reduced with 7% compared to the 1990 level. When we extrapolate the historical trend (fig. 2) to 2000, and assume, that the weight of the car will grow with 1%, then the average car in 2000 only uses 7% less fuel per kilometer, resulting in 9% more fuel consumption for the total fleet, using the 1.7% growth per year.
In fig. 4 the fuel consumption is shown for the total fleet in 2000.

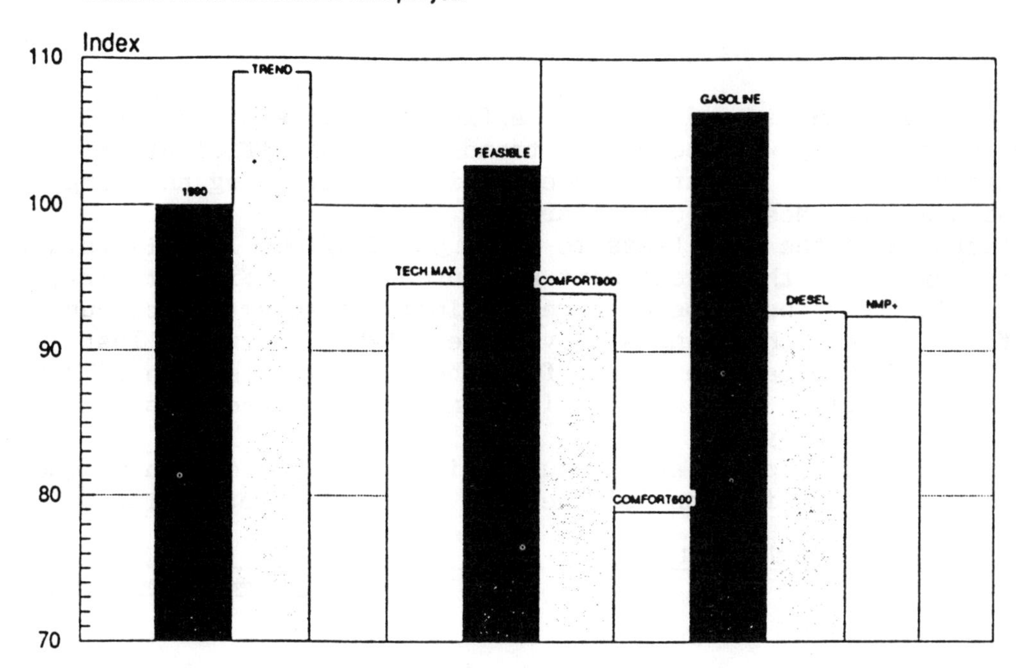

Fig.4 CO_2-emissions total car fleet in The Netherlands in 2000

Depicted is the situation in 1990 which is indexed on 100%, the situation, when assuming the technical improvements will follow the extrapolated historical trend (TREND), and the situation, when using the maximum possible tecqniques, as described above (MAX).
Also is depicted a feasible scenario. The maximum techniques only will be applicated, when there is a market. Creation of this market depends mainly on measurements taken at EC-level, see the last paragraph. Because this will take time, the FEASIBLE scenario seems to be more realistic then the MAX scenario.
In fig.4 the effect is shown when in the coming decade the average weight of the fleet will be lowered, resulting in less comfortable cars (average). COMFORT800 gives the results, when the heaviest cars have a weight of 800 kg, COMFORT600 of 600 kg.
The major conclusion, which we can draw, is that the key to lower fuel consumption of vehicle and fleet is introducing the very light car as soon as possible, and to stabilize the amount of driven vehicle kilometers.

The fuel-efficient car as part of a energy-efficient traffic-system

December 1989 the Netherlands hosted a meeting of specialists on the possiblities of a low-emission and energy-efficient traffic and transport system. It appeared that an ecological acceptable traffic system would not result, even with optimistic assumptions on fuel-consumption of cars. The extent of car-traffic should stabilize or even be reduced, no matter the technological advances. This is in contrast with the momentary growth prognosis.

The small, lightweight car as described before fits well into an energy-efficient traffic-system. This very efficient car is less suitable for long distances because of its lower speed and lack of comfort. Essentially the car will be limited to its trongest asset: an extremely versatile individual means of transport. The luxury of the present cars, as far as this is at the expense of fuel-consumption and environment will have disappeared. The efficient car will then be suitable for distances between 10 or 30 to 80 km's. A high quality public transport which is fast and wide-meshed is much more attractive for long distances.

Necessary policy actions

The development and introduction of efficient cars and the stabilisation of car usage do not comply with momentary trends. Also an efficient driving behaviour is necessary. Policy measures are essential in attaining an energy-efficient transport system. Most important are:

- Increase of the fuel taxes to about 500 ECU/1000 l. This is a powerfull incentive for the development and purchase of efficient cars, efficient driving behaviour (speed, acceleration) and selective car-usage. See also fig. 5, where the changes in vehicle weight during the last 20 years are depicted. One can see, that after the two oil-crises in 1973 and 1980, when the fuel prices were high, weight and thus fuel-consumption decreased during several years.
  The harmonisation of excise-duties in the EEC offers an unique possibility to realize the increase of the fuel taxes on short notice;

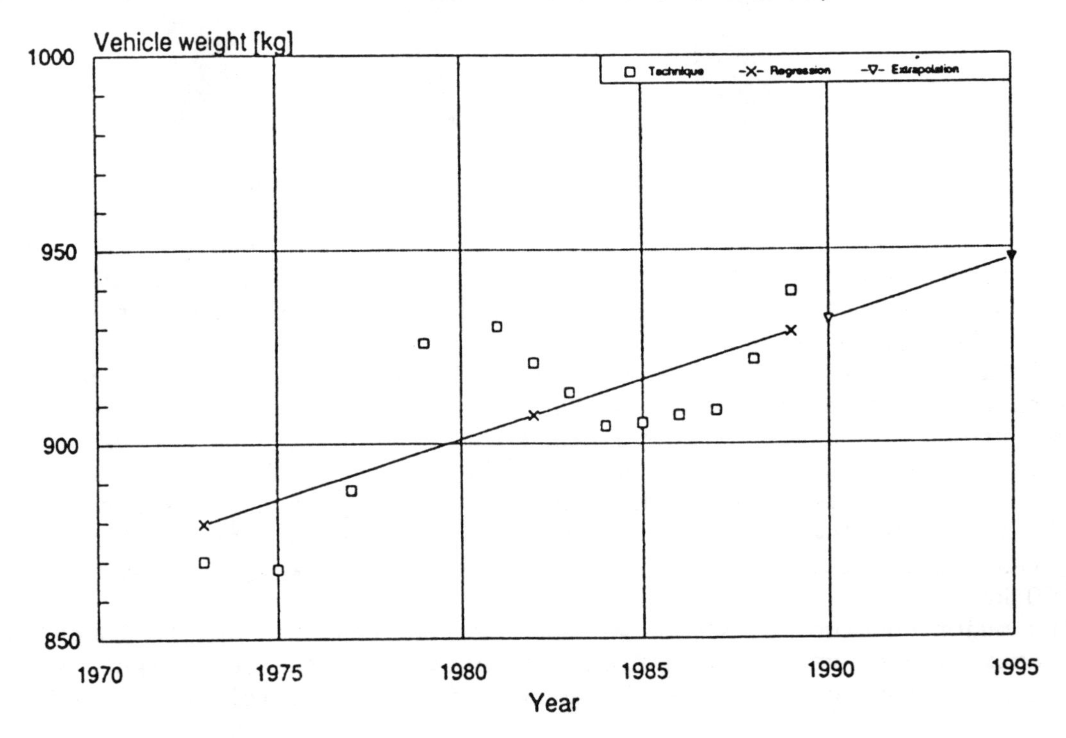

Fig.5 Change in vehicle weight during the last 20 years

- normation of the maximum fuel consumption. Such standards can be used as technology forcing and were implemented in the USA in 1978. The first stage can be a maximum fuel consumption of 8,5 l per 100 km's based on ECE. About 10% of all cars in The Netherlands have a higher ECE fuel-consumption. It may be better to use a for real driving practice more representative driving cycle, e.g. the US-FTP;
- financial incentives should promote the purchase of cars which are more energy-efficient than the standards. The Netherlands used this system succesfully in order to introduce the catalytic converter;
- European CAFE legislation will stimulate production and selling of fuel-efficient cars. It is proven, that such legislation worked well in the US.

- speed limits are necessary to stimulate efficient driving behaviour and to discourage purchase of fast and inefficient cars.

Studies

- Meeting: "Emission-limits and an energy-efficient traffic and transport system" (Dutch);
- European variabilisation of motoring costs (English and Dutch);
- Macro-economic effects of mobility-control (to be published);
- Appraisal of the negative external effects of car traffic (Dutch with English summary);
- Clean buses (Dutch with English summary);
- Traffic and environment in Western Europe (Dutch);
- Regenarative braking systems in city traffic (Dutch);
- City bus 2000, autonomous systems; comparison of different transmissions and regenarative braking systems with regard to emissions and fuel consumption (Dutch with English summary);
- Short-term technical measures for the reduction of CO2-emissions by cars (to be published);
- Small-scale electricity storage (to be published).

REDUCING FUEL CONSUMPTION AND EMISSIONS FROM THE VEHICLE PARK

by J.M. DUNNE, Head, Vehicle Emissions Group
Warren Spring Laboratory, United Kingdom

CO_2 EMISSIONS/FUEL CONSUMPTION REDUCTION

The main aim in reducing CO_2 emissions must be to directly or indirectly reduce the consumption of fossil fuel burning. With respect to petroleum products and specifically that portion used in road vehicles, a number of scenarios need to be considered:

1. Short Term Policies (0-20 years)

General - Establishment of an integrated traffic management policy

The growth of the private motor car and personal freedom to travel independently has directly resulted in stagnation of public transport to the extent that the car is now established as an essential commodity. Reversing this process by investment and incentives could result in substantial reductions in CO_2.

Fiscal incentives - Fuel pricing, vehicle taxation, road taxation, alternative transport incentives, public transport incentives

Fiscal taxes applied to the private motorist can encourage the uptake of smaller and more fuel efficient vehicles, but without tackling public transport infrastructure the gains achieved are likely to be short-lived, the retail price index will be adversely affected and inflation fuelled.

Legal incentives - CO_2 emissions standards, (type approval and in-service controls)

Emissions standards have proven to be an effective method to force technology - provided that its introduction and subsequent reductions in limit levels are made progressively and with due regard to the economic distortions that could arise. Type approval regulations should also be backed up with effective in-service controls using appropriate on-board technology to facilitate enforcement.

Vehicle technology - Power to weight ratio improvements, electronic vehicle and engine and driver management, emissions control trade-off's, reduced drag factors, tyre and transmission losses, improved engine thermal and mechanical efficiencies, improved cold start warm up times, improved in-service durability and control, fuel technology including alternative fuels, "urban" vehicles.

Substantial reductions in CO_2 are possible in the short-term by technological improvements to the standard vehicle, as listed above. Many, particularly driver controls, could encourage the switch to lower powered (cheaper) vehicles without the drawbacks of fiscal price controls, thus producing a double benefit.

Traffic Management - Priority for public transport systems, enforced speed restrictions, improved traffic flow rates, limited acceleration rates, urban restrictions coupled with improved and integrated public transport systems, rail and water development for freight transportation.

Urban traffic management and comprehensive urban public transport systems could remove a substantial proportion of commuter traffic from the public highway. Direct and indirect reductions in CO_2 would result with lower traffic volume and improved traffic flow.

2. Long Term Policies (> 20 years)

Long term policies are not being considered here in detail, however I believe that in general terms it will become essential to switch to alternative means of transportation based on "clean" fuel such as non-fossil fuel derived electricity or hydrogen, bearing in mind that the latter will require large electrical power generation in its own right. Short term policies must be considered together with the long-term aims to accomplish a smooth transition to a totally different transport infrastructure on political, environmental and economic grounds.

REDUCTION OF OTHER GASEOUS POLLUTANTS

Measures to reduce CO_2 can conflict with the aims of reducing other pollutants, particularly NOx. Three way catalyst technology for instance is highly effective in reducing currently regulated tailpipe emissions, but can result in CO_2 increase of in excess of 20%.

Should the need be proven to reduce CO_2 emissions dramatically, then serious consideration should be given to either freezing or even relaxing, at least regionally, the measures being discussed by the EC for introduction by 1993 to permit development of alternative power units, such as lean-burn and direct injection diesel engines.

PART III

TECHNOLOGICAL CONSIDERATIONS

TECHNOLOGY IMPROVEMENTS TO INCREASE FUEL ECONOMY

K.G. Duleep
Energy and Environmental Analysis

Overview

The current concerns over greenhouse gases and global warming has renewed interest in automotive fuel economy. Many conservation groups have called for imposing fuel economy or CO_2 emission standards that would decrease fuel consumption from the current range of 8 to 9 1/100 km in many OECD countries to less than 5 1/100 km by 2005. Such a standard is by no mean physically impossible as cars have been built to attain fuel consumption as low as 2.5 1/100 km. However, it is clear that such low fuel consumption cars do not have the size, performance and luxury features demanded by many consumers, and often do not meet emission and safety standards prescribed in many OECD countries.

No forecast of fuel economy or estimate of fuel economy potential is possible without specific consideration of vehicle attributes. At minimum, consumers value interior room, acceleration performance, specific luxury features such as power steering, automatic transmission and air conditioning, and safety features such as collision resistance, anti-lock brakes and impact protection devices. In a broader context, cars also convey prestige and status to the user, but these are more difficult variables to consider in estimating fuel economy. Other attributes may be driven by regulation, such as emission standards. High fuel economy (or low fuel consumption) values are possible by relaxing the demand for attributes, but this is not a technology question. Rather, it must rely on forces to shift demand by consumers. OECD countries have largely resorted to tax on cars by size and tax on fuels to shift demand,

and no other options are readily available that do not impinge on consumer freedom to choose.

EEA has recently undertaken analysis for the U.S. Department of Energy to estimate future fuel economy potential to 2001 for domestically produced cars. One assumption used in the study was to estimate a maximum fuel economy (MPG) level possible holding attributes constant at 1987 levels for the new car fleet in terms of size, performance and luxury options. A second assumption was to estimate the potential changes to attributes based on information on manufacturer product plans and current demand trends. Increasing affluence in the U.S., as well as low fuel prices relative to the early 1980's, has led to consumers demanding larger, more powerful and more luxurious cars. Such as trend has been noted in almost all OECD countries.

Analysis Methodology

In order to forecast future fuel economy of individual vehicles or vehicle types, EEA has developed a methodology that relies on three steps:

- Specification of a baseline of fuel economy and vehicle technology attributes.
- Identification of individual fuel economy technologies applicable to each vehicle type and fuel economy benefits associated with these technological improvements.
- Adoption of technologies as dictated by synergistic and non-additivity constraints to attain different levels of fuel economy.

The methodology was developed by EEA over the last 10 years under contract to the DOE, and has received extensive public review. A key aspect of the methodology is the assumption that a vehicle's fuel economy can be estimated with reasonable accuracy if its attributes such as weight and performance (horsepower) are known, and the technologies embodied in the engine and transmission are identified. The implication of this view of fuel economy is that different manufacturers building similar cars with similar technologies can reach nearly identical fuel economy levels in a given market segment. In fact, examination of the current car models available and their fuel economy levels suggest that, for a given level of technology, fuel economy within market segments is almost independent of manufacturer. Of course, some manufacturers have introduced more technology into the market place than others and, hence, are ahead in fuel economy.

The baseline used for forecasting was derived from actual 1987 sales data and individual vehicle fuel economy and weight by engine/transmission type. In addition, information on engine types, valves per cylinder, type of fuel injection, presence or absence of roller cam followers, specific output distribution, etc., were compiled and used for the forecast baseline. Individual technological benefits applicable to each market segment and the resulting forecasts are discussed below.

Technologies

The automobile and its' motive power source, the internal combustion engine, have been in existence for a century. As a result, the technology of automobiles is well understood and mature. Improvements are of the evolutionary type and are rarely revolutionary so that future technological improvements can be anticipated. Another factor aiding in the identification of technologies that can be commercialized over the next decade is the long lead time associated with technology development, design, tooling and production. In fact, technologies that can be commercialized to 2001 must already be beyond the "proof-of-concept" phase to satisfy the lead time constraints for introduction into the marketplace.

An extensive list of technologies has already been compiled by EEA, and their fuel economy benefits estimated from a combination of sources including research papers, actual benefits from a few vehicles already featuring this technology, and from manufacturer submissions to DOT. More recently, EEA has published two major reports, one for the Office of Technology Assessment,[1] and the second for the U.S. Department of Energy.[2] These reports have attracted considerable attention from the auto-manufacturers, and the technology benefit estimates extensively critiqued. EEA has also publicly defended the estimates, and manufacturer inputs have been examined and taken into account in revisions to these estimates.

A comprehensive list of technologies that can be used to improve fuel economy over the next 10 years is provided in Table 1. Technology costs and benefits are directly dependent on what a particular technology is being compared against. For example, the benefit of multi-point fuel injection is dependent on whether the comparison is against a carburetor or against throttle-body fuel injection. The comparison baseline for each technology is documented in Table 1.

A second factor that has caused some misunderstanding in public reviews is a semantic issue, when a specific technology name designation represents a group of technologies. The introduction of certain technologies allow other changes to be made to a vehicle that contribute to a greater change in fuel economy than if it were introduced without other changes. EEA's technology name designations include an entire group of technological changes for estimating technology benefits for certain technologies. For example, EEA's "Front Wheel Drive" technology benefit is based on the fact that the transverse engine location allows considerable exterior size reduction and accompanying weight reduction while keeping interior room constant. The secondary effects include engine downsizing to keep performance constant. Thus, the benefits associated with the term "Front-Wheel Drive" <u>includes</u> all of these primary and secondary effects. Similarly, the benefits for "overhead cam" engines is based on the fact that existing overhead valve engines have low specific output, and a new overhead cam engine of <u>lower</u> displacement can be used to replace the older engine while providing equal performance. The central variables that are <u>maintained constant over time are the interior volume of the car as well as the performance</u> as defined by its acceleration capability. These issues are of importance since it is the most common source of misunderstanding of the EEA technology benefit estimates.

TABLE 1

TECHNOLOGY DEFINITIONS

Technology		Fuel Economy Baseline
Front Wheel Drive	-	Benefits include effect of weight reduction and engine size reduction starting from a late-1970's vintage rear-wheel drive design.
Drag Reduction I	-	Based on C_D decreasing from 0.375 in 1987 to 0.335 in 1995, on average.
Drag Reduction II	-	Based on C_D decreasing from 0.335 to 0.30 in 2001, on average.
Torque Converter Lock-up	-	Lock-up in gear 2-3-4 compared to open converter.
4-Speed Auto Transmission	-	3-speed auto transmission at same performance level.
Electronic Transmission Control	-	Over hydraulic system, with control of shift schedule and lock-up of torque converter.
Accessory Improvements	-	Improvements to power steering pump, alternator, and water pump
Lubricants	-	5W-30 replacing 10W-40 oil, plus synthetic axle lubricants.

TABLE 1

TECHNOLOGY DEFINITIONS

(Continued)

Technology		Fuel Economy Baseline
Overhead Camshaft	-	OHV engine of 40 BHP/liter replaced by OHC engine of 50 BHP/liter but with smaller displacement for constant performance.
Roller Cam Followers	-	Over sliding contact follower.
Low Friction Pistons/Rings	-	Over 1987 base except for select engines (already incorporating improvement).
Throttle Body Fuel Injection	-	Over carburetor (includes air pump elimination effect).
Multi-Point Fuel Injection	-	Over TBFI. Includes effect of tuned intake manifold, sequential injection and reduced axle ratio for constant performance.
4-Valve Engine (OHC/DOHC)	-	Over two-valve OHC engine of equal power. Includes effect of displacement reduction and compression ratio increase.
Tires	-	Over 1987 tires, due to improved construction.
Intake Valve Control	-	Lift and Phase Control for intake valves, at constant engine output.

Separately, diesel engines and two stroke engines were not considered in this analysis, principally because it appears that a 0.4 g/mi NO_x standard is virtually certain in the U.S. for the mid-1990's. Neither engine has demonstrated a capability to meet this standard, and was deemed too risky for inclusion in the study. Such technologies may be useful in other OECD countries with less stringent standards.

Using engineering principles, it is relatively straightforward to derive an equation that relates fuel consumption (the inverse of MPG) to vehicle attributes such as weight, drag, engine size and efficiency. Using the notation employed by Sovran,[3] fuel consumption over the FTP driving cycle, g (in gallons) is given by the following equation:

$$g = \frac{\overline{bsfc}}{N_d} E_{TR} + \overline{bsfc}\, E_{AC} + G_o \left[T_{ID} + T_{Br}\right]$$

where: $\overline{bsfc}$ is the average specific fuel consumption over the cycle

N_d is the drivetrain efficiency

E_{TR} is the energy consumed in overcoming tractive forces

E_{AC} is the energy consumed by accessories

G_O is idle fuel consumption rate

T_{ID}, T_{BR} are the time at idle and during braking with the throttle closed.

In addition, E_{TR} can be written as the sum of energy consumed for overcoming inertia force (I), aerodynamic drag (A) and tire rolling resistance (R).

i.e. $E_{TR} = E_I + E_A + E_R$

As can be seen from the above equations the fuel consumption benefits from weight reduction, drag reduction, the rolling resistance reduction, idle fuel consumption and accessory power reduction do not have negative synergies with each other. In the sense that the engine size itself does not have a large impact on bsfc, improvements in the above factors do not have synergy with each other or engine bsfc, as a first order approximation. In other words, synergistic effects are an order of magnitude smaller than primary non-synergistic effects.

Engine efficiency itself is a function of three independent factors:

- Thermodynamic efficiency as dictated by compression ratio and combustion efficiency
- Pumping loss which is dependent on the load factor and intake design
- Internal friction which is dependent on engine mechanical design

The drivetrain efficiency is already quite high, but the drivetrain type and the gear ratios strongly affect pumping loss. An examination of the list of technologies shows that the only significant synergy lies in technologies that contribute to pumping loss reduction.

Technology non-additivity refers to the fact that many technologies listed in Table 1 are mutually exclusive, e.g. throttle-body fuel injection and multi-point fuel injection. These non-additivity constraints are simply handled in the forecast by tracking non-additive technology market penetration. Since two non-additive technologies cannot be present in the same vehicle, the sum of the two technologies' market penetrations can never exceed 100 percent.

Specific synergies considered in this model are for intake valve control and five-speed automatic transmissions or six speed manual transmissions. Both valve control and transmission technologies reduce pumping loss, and hence, if adopted in the same car do not produce the same benefit as the sum of benefits acting separately. For these technologies the total benefit was reduced from 8.5 percent (without synergy) to 6.5 percent.

Forecasts

Forecasts of fuel economy for the U.S. domestic new car fleet (GM. Ford and Chrysler) were derived for 1995 and 2001. For each year, two scenarios have been developed. The first is a product plan based scenario that attempts to forecast what will happen in the absence of new fuel economy regulations. The second is a "maximum technology" scenario that utilizes all possible technologies regardless of cost-effectiveness. EEA analysis has revealed that technologies that are available and cost effective are generally adopted by manufacturers as part of the product plan. Here, the term "cost-effectiveness" is based on the incremental price of the technology being lower than the cost of fuel saved over 4 years (50,000 miles for cars and 60,000 miles for trucks). In addition, product plans are often revealed in the trade press and are tracked by EEA to the extent available. The cost-effectiveness criterion is used to estimate if a technology will be included in the product plan at a specified fuel price.

All of the technology analysis is based on maintaining constant interior volume and constant performance. A review of the product plans and actual events between 1987 and 1990 show that this assumption does not hold as cars are increasing in size and offering improved performance. New products in the future are expected to continue these trends. However, the simultaneous introduction of new technology often results in no net reduction (or even an improvement) in new car fuel economy even after weight increases and performance enhancements have been accounted for.

One aspect to be considered for the 1995 model year is the lead time requirement. The 1995 model year is now only about 4 1/2 years away, and most of the products have already been designed for this date. As a result, the 1995 analysis includes all of the known increases in performance, size and luxury that have already occurred since 1988 or will occur shortly, since these cannot now be changed. The lead time constraints are also instrumental in preventing any significant increases in fuel economy by 1995. Manufacturers can make marginal improvements over those already planned for 1995 by accelerating (or "pulling ahead") technologies planned for introduction in 1996/97 and by extending the market penetration of technologies that are used in parts (but not all) of its product line. These are not considered here because the effect of pulling ahead technologies is small.

For 2001, such restrictions are less applicable. The product plan case holds attributes constant from 1995 while maximum technology scenario holds attributes constant from 1987.

The maximum technology scenario considered in this analysis is, indeed, a real maximum. The scenario envisages that all products are redesigned employing new materials such as aluminum and advanced plastics extensively, and that the only engines sold are multi-valve small displacement engines with intake valve control in virtually all but the smallest cars. The extent of such changes would impose a heavy burden of refueling for the industry and would require unprecedented and risky changes to every product sold. Moreover, many technologies introduced under this scenario will never pay for themselves in terms of fuel savings to the consumer. These factors should be understood when considering the maximum technology scenario results.

Results

The estimates for 1995 fuel economy is shown in Table 2. Unfortunately, of the anticipated technological benefit of 16 percent, nearly half will be lost to increased consumer demand for size, luxury and performance. As a result, the fleet would have attained 31 MPG by 1995 if consumer demand had remained at 1987 levels, but will actually be about 29 MPG. Table 3 shows the potential to 2001. Again, under the product plan, fuel economy is expected to

TABLE 2

U.S. DOMESTIC MANUFACTURERS

1995 PRODUCT PLAN CASE

Technology	F/E Gain (%)	Penetration Increase from 1987 (%)	1995 Penetration	Fleet F/E Gain
Front Wheel Drive	10.0	12	86	1.20
Drag Reduction (C_D ~0.33)	2.3	80	~100	1.84
4-speed Automatic Transmission	4.5	40	80	1.80
Torque Converter Lock-up	3.0	20	90	0.60
Electronic Transmission Control	1.5	80	80	1.20
Accessory Improvements	1.0	80	N/M	0.80
Lubricant/Tire Improvements	1.0	100	100	1.00
Engine Improvements				
- Overhead Camshaft	6.0	45	69	2.70
- Roller Cam Followers	1.5	40	95	0.60
- Low Friction Pistons/Rings	2.0	80	100	1.60
- Throttle Body FI	3.0	12	40	0.36
- Multi-point FI (over throttle body)	3.0	12	60	0.36
4 valves per cylinder engine				
- 4 cylinder replacing 6*	10.0	18.0	18	1.80
- 6 cylinder replacing 8*	10.0	12.0	12	1.20
Effect of safety standards				-1.00
Total F/E Benefit				16.06

1987 Fuel Economy: 26.7 mpg

1995 Fuel Economy: 31.0 MPG** (at constant size/performance)
29.0 MPG (at expected 1995 levels of size/performance

* 1987 distribution: 20.5% V-8, 29.5% V-6, 50% 4 cylinder

TABLE 3
POTENTIAL FUEL ECONOMY IN 2001 UNDER ALTERNATIVE SCENARIOS

	Fuel Economy Benefit	Product Plan		Max. Technology	
		Mkt. Pen 1995 - 2001	F/E	Mkt. Pen 1995 - 2001	F/E
Weight Reduction	3.3/6.6	100	3.30	100	5.28
Drag Reduction	1.15/2.3	80	0.92	80	1.84
Intake Valve Control	6.0	30	1.80	70	4.20
Overhead Cam Engines	6.0	30	1.80	30	1.80
6 cyl./4V replacing 8 cyl.	10.0	4	0.40	8	0.80
4 cyl./4V replacing 6 cyl.	10.0	6	0.60	12	1.20
4 cyl./4V replacing 4 cyl.	5.0	20	0.50	50	2.50
Multi-point fuel injection (over TBI)	3.0	40	1.20	40	1.20
Front wheel drive	10.0	5	0.50	13	1.30
5-speed auto. transmission*	2.5	20	0.50	40	1.00
Continuously variable transmission*	2.5	20	0.50	40	1.00
Advanced engine friction reduction	2.0	100	2.00	100	2.00
Electric Power Steering	1.0	5	0.05	30	0.30
Tire Improvements	0.5	20	0.10	100	0.50
Total F/E Benefit (%)			13.27**		23.48**
Unadjusted CAFE (MPG)			32.85		38.3

Note: Product plan scenario starts from a different 1995 base than the maximum technology scenario

* Over 4-speed auto transmission with lock-up.

** Accounts for synergy between advanced transmissions and intake valve control.

increase to 33 MPG starting at 29 MPG in 1995. A maximum technology case holding product attributes constant at 1987 (baseline) levels indicates that 38.3 MPG will be possible. However, the mechanism for holding attributes constant is yet undefined. It should be also noted that these projections <u>do not consider</u> the effect new safety and emissions regulations that have been enacted or are under consideration. Such new standards could result in an additional 1 MPG loss in 1995 and 2001.

These results will change for other OECD countries. Most OECD nations outside of the U.S. and Canada already utilize small displacement, overhead cam engines and further engine downsizing will be problematic. In general we anticipate that other nations cannot expect even the indicated benefits at constant attributes. On the other hand, many European vehicles seem poorly optimized for fuel economy due to the emphasis on performance. Thus, the fuel economy potential from performance reduction is much larger than in the U.S.

It is, indeed, difficult to break the current consumer demand trends for increased size, luxury and performance in a free society. While technology can provide benefits, it cannot be used to attain the 100 MPG (2.5 1/100 km) that some prototypes have demonstrated without somehow addressing consumer requirements. One possibility is to allow consumers to purchase a high fuel consumption "status" car but provide cheap urban commuter vehicles (that can have very low fuel consumption) on a rentable basis for urban commuting. Another would be an extensive campaign to change demand trends, like those used to counter smoking. In any event, it is clear that more innovative solutions will be required, rather than relying on technology alone to solve greenhouse problems.

REFERENCES

1. EEA "Developments in the Fuel Economy of Light-Duty Highway Vehicles" report to the Office of Technology Assessment, September 1988.

2. EEA "Domestic Manufacturers Fuel Economy Capability to 2001 - An Update" Report to the DOE, October 1989.

3. Sovran, "Tractive - Energy Based Formulae --" SAE Paper 830034, February 1983.

REDUCTION OF AUTOMOBILE FUEL CONSUMPTION.

Jon R. Bang
National Institute of Technology, Norway

The fuel consumption of a car is given by its:

1) Weight
2) Air and rolling resistance
3) Engine efficiency
4) Transmission efficiency and gearing ratios

Reduced consumption may be achieved by a change in only one of the parameters above, but a better way is to reduce both 1) and 2) and thoroughly optimize the drive-train to the new vehicle concept.

1) Some 15 - 20 years ago many passenger cars weighted not more than 700-800 kg. Today the most common weight is 1100 to 1200 kg.
   The last decade the cars has increased in weight. As an example an Audi 80 from 1975 had a curb weight of 860 kg. A 1990-model Audi 80 weights close to 1100 kg.

   As a thumb rule for a medium size car, a 100 kg weight reduction may give 0.3 liter/100 km or ca.4% lower fuel consumption.

   Despite the common increase in weight, the general trend in fuel consumption is downward. From 1980 to 1988- passenger car models reduction in consumption was reported to be 9.9% in Sweden, even if the average weight increased ca. 4%. Their performance (kW/kg) was also increased 12.6% at the same time.

The probable main reasons for the weight increase is a general demand for more comfort (with a lot of servoes, electronic circuits, wiring and sophisticated options) and safety constructions and details.

Within a relative short time it should be technically possible to construct and produce a passenger car with a curb weight of 6-700 kg. Extensive use of aluminum and plastic could reduce the weight by 20% compared to conventional steel. A general downsize would be a normal part of such a development. With a normally good matched engine with today's efficiency, the fuel consumption could then be reduced by 25-30% . Several manufacturers presented such research vehicle already 10 years ago. The technique is not the main problem here, but the market and business acceptance.

2) Contrary to the weight, the air resistance has been reduced dramatically the last ten year from typically 0.45 to 0.30-0.34 today. Values around 0.30 should be within the technical reachable for most car manufacturers within some years. Down to 0.22 - 0.25 may be reached in a decade for some models, but hardly lower. These values may be a limit for a reasonable practical passenger cars.
The average fuel consumption drops ca. 3-4% if the air resistance is reduced by 10%. Going from typical C_W=0.34 today to 0.30 in some years, would then give ca. 5% reduced consumption. This is not so much. Weight reduction might be a more fruitful way to go.

The rolling resistance has also been reduced, from typically 0.14 (rolling resistance coefficient) some year ago, to 0.12-0.13 today. Even lower values will be reached in the future. However the impact of this improvements on the fuel consumption are not dramatic, but more a contribution to a step by step reduction.

3) Even though the gasoline engine (otto) has improved its efficiency very much the last decade, it can not compete with a turbocharged direct injected diesel engine. Small direct injected diesel engines, as developed today for passenger cars, have 20-25% lower consumption than their gasoline alternatives. Direct injected methanol or gasoline engines may be on the production line within a decade, and their efficiency will probably be on the same level. Their combustion cycle will be more a diesel-cycle with high compression ratio and reduced pumping losses together with reduced heat loss. Their energy efficiency may be some 30% better than modern gasoline engine today. They might be lighter and smaller and thus contribute to a vehicle down-scale.
Methanol fuel has potentials to reduce energy consumption and emissions of greenhouse gases. Methanol's high octane number allows a high compression ratio and thereby increased engine efficiency which gives less CO_2. The high heat of evaporation for methanol gives the opportunity for a higher engine output from a given engine volume. Given a specific engine power, a methanol engine could then be smaler and lighter. A lighter engine means a lighter vehicle, which then again gives reduced energy consumption.

The power/weight ratio of many modern passenger cars are today fare beyond any rational needs. They are for pleasure. It can be shown that a 30 kW engine should be sufficient for normal use of a passenger car with 800 kg curb weight. Use of smaller engines would allow them to work with higher specific output (kW/l) with better fuel burning efficiency.

4) It is doubtful if the efficiency of modern step transmissions could be increased very much. Modern 4- or 5-speed automatic transmission with lock-up may be close to what can be reached. Continuously variable transmissions may improve fuel consumption by enabling the engine to run at a more efficient speed for a given power output. However the experiences with the very few models today with such transmissions, have not meet the expectations too much.

Development of more fuel efficient transmissions may give some, but we do not think the biggest contribution to consumption and emission reduction.

The <u>emissions</u> are a product of the exhaust flow rate and the concentrations of the pollutants. For gasoline cars the flow rate is fairly proportional to the fuel flow rate, ca. 16 kg exhaust/kg fuel. For diesels the exhaust flow rate is closely proportional to the engine speed.

There is now a trend towards a common public acceptance of the catalyst technique. With this technique the concentrations of the pollutants depends of how extensive the car manufacturer has used the techniques potential. However we do not think manufactures are going to produce minimum emission cars without regulatory pressure or incentives.

For cars driven on fossil fuels there might be a a minimum for emissions which could hardly be beaten. This limit is close to the California "Low Emission Vehicle" limits: CO: 2.1, NMHC: 0.047, NOx: 0.13 g/km. With other fuels the CO and NMHC limits could be lowered.

Possibilities for Energy Saving and Reduction of Exhaust Emission in Motor Vehicles

Dr. Kazuo Kontani

Mechanical Engineering Laboratory
Agency of Industrial Science and Technology
Ministry of International Trade and Industry, Japan

The most prompt and directly effective option among the various ways of reducing energy consumption of motor vehicles would be a transition toward smaller cars. Since this would also reduce the absolute quantity of exhaust emission, it would also be effective on the exhaust gas problem. As regards to the possibilities for immediate improvement of the fuel consumption in motor vehicles, the followings can be formed.

Without sacrificing performance, about 3-5 %.
Sacrificing performance to some extent, about 20 %.
By shifting to liter (or around 1300 cc) cars, about 50 %.

These estimates are relative to the totality of gasoline-powered passenger cars currently operating in Japan, and are subject to the assumption that the current regulations on exhaust gas levels will be maintained.

As an example of corroborative data in this connection, Figure 1 shows the relation between the weight of motor vehicles in Japan and their 10-mode fuel consumption. The abscissa represents the inertia weight of the vehicle, while the ordinates show the fuel consumption (km/l) for manual and automatic transmission vehicles separately, as well as the volume-weighted averages for all vehicles. The bar graph in the figure shows the number of units at each weight class in %.

On the somewhat gross assumption that a nearly complete shift to liter cars (inertia weight of around 750 kg) were accomplished, fuel consumption would then be improved by 34% for M/T vehicles and 48 % for A/T vehicles. Also, the fuel consumption (km/l) of M/T cars exceeds that of A/T by an average of 35.5 %, and by 12-29 % if only cars in the same weight class are compared. This difference can be regarded, in a sense, as a gap which presumably

In addition, the other main technical factors with major potential for future energy saving and exhaust gas reduction in motor vehicles are as follows.

Lean combustion: This reduces heat and pumping losses and approaches the air cycle, thereby improving thermal efficiency. The development of catalysts for lean burning constitutes a key in this respect.

Stratified charge engines: This realizes lean combustion in the overall average, and has been attempted for many decades, but with no decisive results.

Two stroke engines; The potential advantages of two stroke engines are being re-examined in the context of current progress in electronic control technology. This offers the possibility of reduced weight and size, flat torque and improved fuel economy under partial load conditions.

Improvement of transmission: Possibilities include increasing the number of stages or continuously variable transmission (CVT). CVT could be more efficient, in some cases, than manual transmission.

Recovery of braking energy: Possibilities lie in oil-pressure acumulation or combination of flywheel and CVT, etc.

Catalysts for Diesel engines: The development of catalysts for deoxidization of NO_x would give considerable progress in the problem of Diesel exhaust emission, which would thereby create additional possibilities for improvement of fuel consumption.

The simultaneous achievement of low fuel consumption and low environmental pollution is not an easy target. Efforts to reduce fuel consumption have already been pursued for many years, and demands for the still further reduction of fuel consumption would be difficult to satisfy by mere technological approaches alone. Retrogression with respect to the objective of low environmental pollution should not be permissible, however, if energy saving is the primary consideration at this moment, then further strengthening of restrictions on exhaust emissions should presumably be avoided at the present time.

could be filled by appropriate technical improvements.

Of course, a transition toward smaller and lighter vehicles would have a considerable industrial and economic impact, requires consideration from various standpoints, and would be difficult to implement immediately, but nevertheless constitutes the most effective method of coping with the issues of energy conservation and environmental protection. However, the current users' demand is toward high performance and high grade cars, that is, toward large vehicles (see Figure 2, 3). Leading society in the direction of smaller vehicles is a question charged on government policy, and could be effected by various measures, one of the major factors involved being the price of fuel. To corroborate this one may cite, for example, the fact that the registration of new vehicles in Japan experienced a little stagnation just after the two oil crises.

Next, under the assumption that the present diversity of motor vehicles continues, consider the possible scope of subsequent improvement in fuel consumption. Current gasoline-powered motor vehicles utilize no more than about 10-15 % of the fuel energy under practical operating conditions. On the other hand, the maximum efficiency of a gasoline engine is limited to 30 %, and this is reduced to around 25 % by losses inside vehicle (x 0.9) and outside (x 0.9). Thus the demand to improve fuel consumption of gasoline powered vehicles 30-50 % better than the current level, that is, to get fuel consumption of 15-23 (or even 30) % would be very difficult.

Technological approaches which are capable of contributing to improved fuel consumption are listed in Table 1. Up to the present, measures adopted in the endeavor to reduce the fuel consumption of motor vehicles have included improvement of engine efficiency as well as reduction of vehicle air resistance and weight. Adding to these, hereafter, concomitant with the progress and increasingly wider application of electronic technology, improvement of the power train system, including the transmission, as well as the introduction of "variable-mechanism" technology, will presumably constitute important factors in fuel saving. That is, how to keep the engine in better operating condition will be an essential view point in improving the automobile fuel consumption.

The variable-mechanisms which could provide improvements in fuel consumption are listed in Table 2.

Figure 1 Inertia Weight and 10-mode Fuel Consumption (in 1988, Japan)

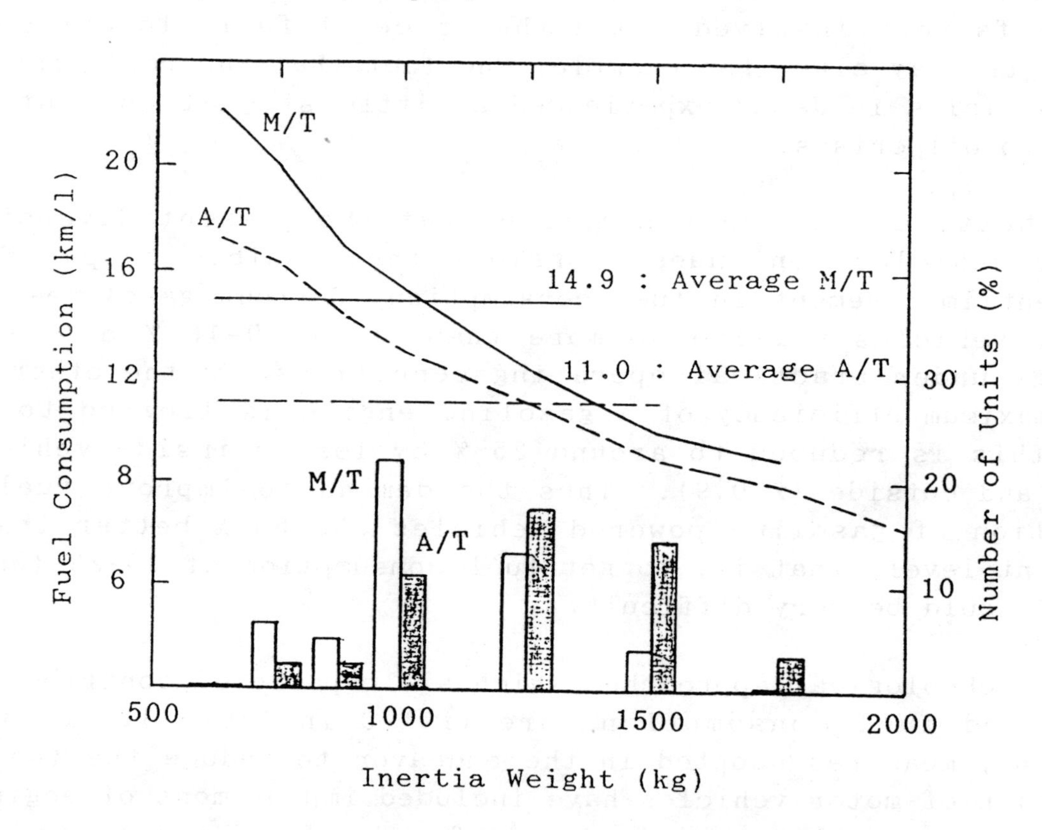

Fig. 2 Increase of High Performance Cars

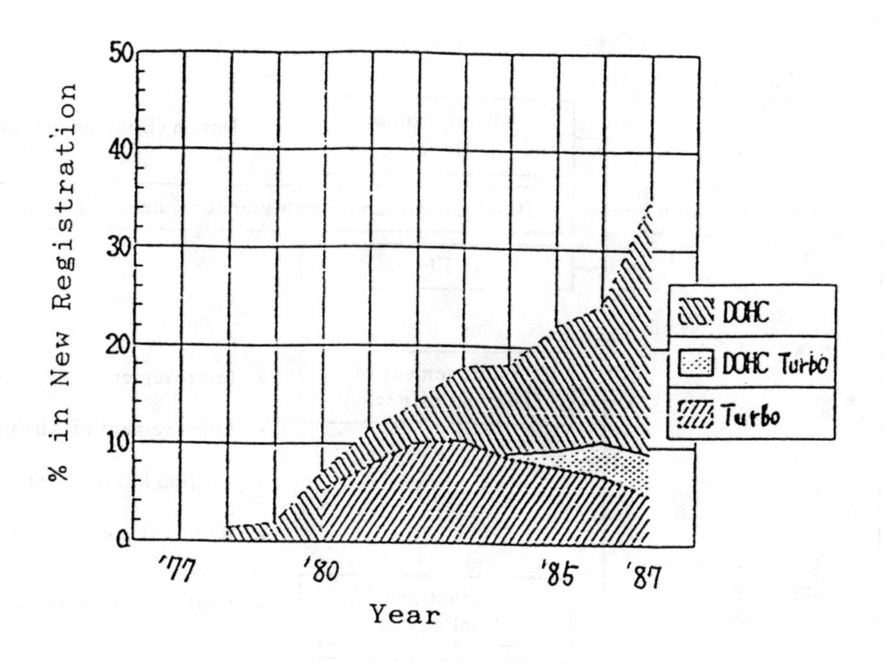

Fig. 3 Increase of larger Cars

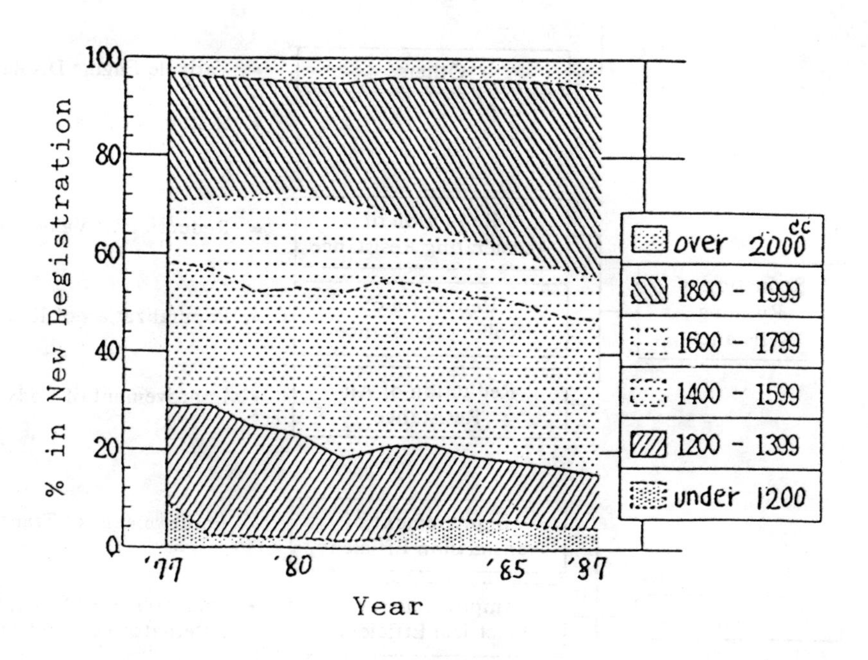

Table 1 Measures of Fuel Efficiency Improvements

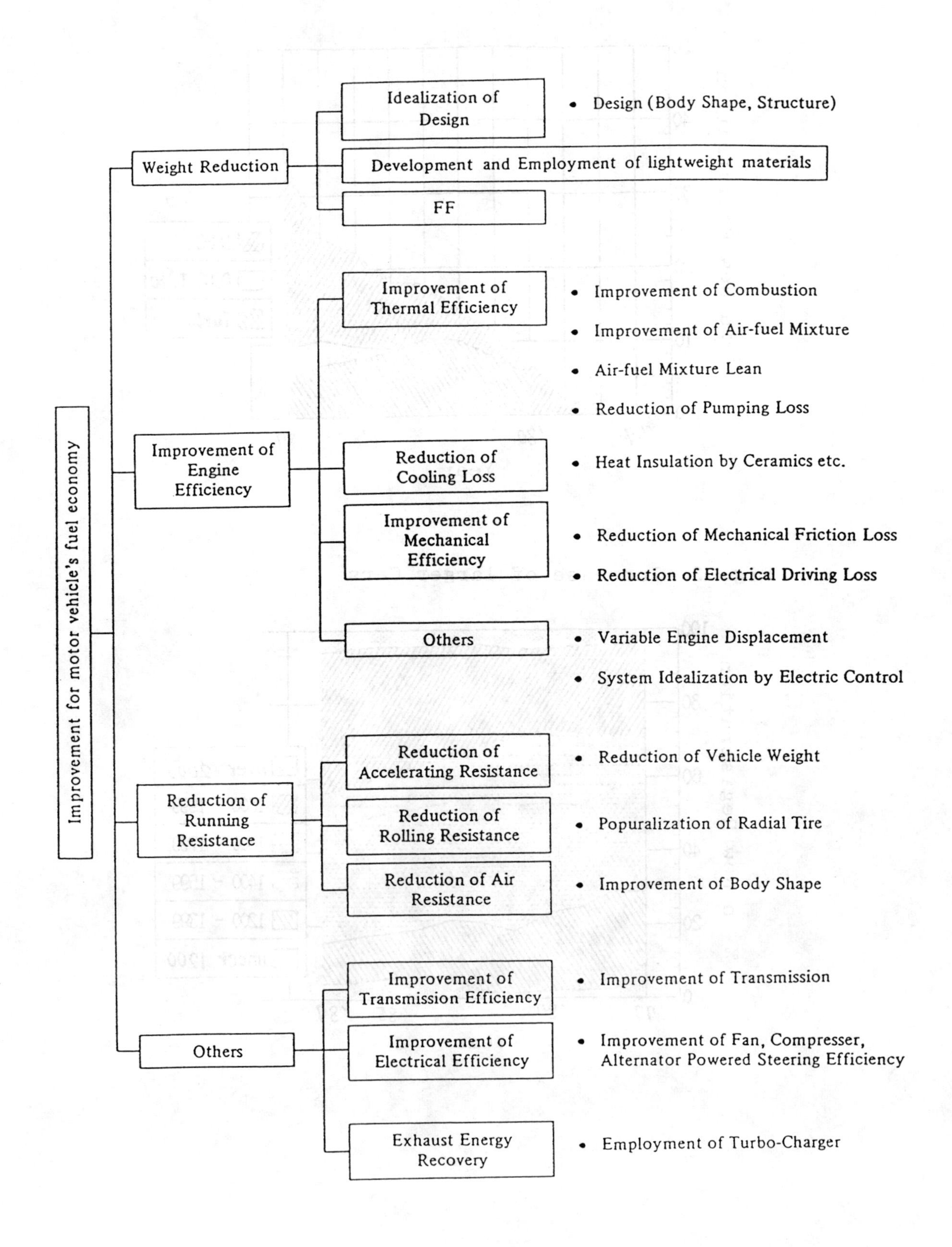

Table 2. Variable-mechanism tecnnologies potentially applicable to reduction of fuel consumption

MECHANISM	EFFECT
(1) Already adopted	
Length of suction pipe	Flattening of torque
Swirl control of intake air	Fuel consuumption and stabilization of idling
Valve timing	Output, fuel con-consumption and idling
Variable turbo-charging	Fuel consumption
Variable number of cylinders	Fuel consumption under partial load
(2) In intermediate stage of development	
Valve timing and lift	Output
Drive by wire	Precise control, improved response
Variable capacity flywheel	Acceleration and fuel consumption
Variable speed drive and/or electrical drive of auxiliary devices	Load reduction
(3) Long-term possibilities	
Variable stroke	Reduction of pumping and friction losses
Variable compression ratio	Output and fuel consumption

OECD/IEA Informal Expert Panel

on

LOW CONSUMPTION LOW EMISSION AUTOMOBILE

ROME, 14-15th February 1990

HOW TO REDUCE FUEL CONSUMPTION OF ROAD VEHICLE

Jean DELSEY

INRETS-LEN (*)

I - INTRODUCTION - PRIORITIES

When you have to move a road vehicle, four resisting forces must be overcome :

- rolling resistance
- aerodynamic resistance
- inertia in period of acceleration
- gravity along a slope.

A rapid analysis shows that the manufacturer's priorities should be to reduce

- aerodynamic drag
- rolling resistance
- weight.

But in fact, the automotive engine has a very poor efficiency and the average distribution of the use of energy is represented by the following graph.

* INRETS - Laboratoire Energie - Nuisances
109 Av. Salvador Allende - 69500 BRON France

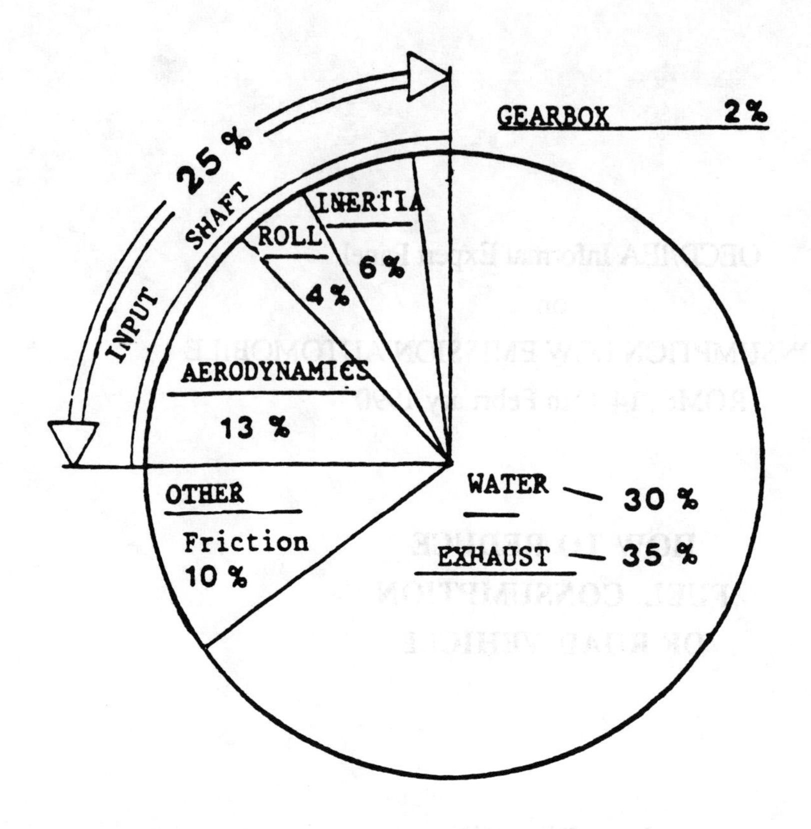

Only 25 % of the total energy is used to move the vehicle (calculated on mixed cycle : urban, road, highway).

So, the priority number one is to increase the efficiency of the propulsion system :

- better combustion
- reducing losses in friction
- optimisation of the temperature of cooling system
- better adaptation of gear boxes.

II - MAJOR PROGRAMS IN FRANCE - RESULTS

To evaluate the ultimate possibilities, the French administrations have decided three programs to build low fuel consumption vehicles :

1979 → 1983 on medium size vehicles
EVE - EVE + (RENAULT)
VERA - VERA + (PEUGEOT)

1980 → 1986 on small size vehicles
VESTA (RENAULT)
ECO 2000 (PEUGEOT)

1980 → 1990 on heavy duty truck (40 tonnes)
VIRAGES (RENAULT V.I.).

The physical caracteristics and the results for the fuel economy are described on the following tables.

	R18	EVE	EVE +
Weight	920	845	845
Cx	0,39	0,24	0,225
SCx	0,73	0,47	0,44
Engine	gasoline	gasoline	Diesel direct injection turbocharged
	1400 cm3 46 KW	1100 cm3 39 KW	1600 cm3 37 KW
Max. speed	150 km/h	157 km/h	165 km/h
Fuel Consumption			
90 km/h	6,3	4,1	3,5
120 km/h	8,4	5,5	4,4
Urban	9,4	6,6	5,8
Average	8,0	5,4	4,6

Low Fuel Consumption for Medium Size Vehicles from RENAULT.
Comparison with Commercial Model R18 - Reduction of 43 % for EVE +

	305	VERA	VERA +
Weight	940	752	700
Cx	0,44	0,30	0,28
SCx	0,78	0,57	0,47
Fuel Consumption			
90 km/h	6,2	4,2	3,5
120 km/h	8,4	5,6	5
Urban	8,9	6,3	5,2
Average	7,8	5,4	4,6

Low Fuel Consumption for Medium Size Vehicles from PEUGEOT.
Comparison with Commercial Model 305 - Reduction of 41 % for VERA +

	RENAULT 5 SL	VESTA
Weight	725 Kg	473 kg
Cx	O,34	0,19
SCx	0,60	0,30
Engine - gasoline	1108 cm3	716 cm3
	34 KW	20 KW
Max. speed	143 km/h	140 km/h
Fuel Consumption	5 vitesses	5 vitesses
90 km/h	4,1	2,05
120 km/h	5,6	2,73
Urban	5,8	3,66
Average	5,2	2,8

Low Fuel Consumption for Small Size Vehicles from RENAULT.
Comparison with Commercial Model R5 SL - Reduction of 46 % for VESTA

	ECO 2000 Gasoline	ECO 2000 Diesel
Weight	458 kg	510 kgs
Cx	0,235	-
SCx	0,38	-
Engine	750 cm3	903 cm3
		Direct injection-turbocharged
	27 Kw	29 KW
	5 vitesses	5 vitesses
Max speed	151 km/h	163 km/h
Fuel Consumption		
90 km/h	2,3	2,1
120 km/h	3,2	3,1
Urban	3,5	3,2
Average	3,0	2,8

Low Fuel Consumption for Small Size Vehicles from PEUGEOT.
Two versions, Gasoline and Diesel.

The results of the programs on the commercial production were very positive as we can see for the evolution of fuel consumption on small size vehicles.

Values are indicated for PEUGEOT 104 where only the engine was modified between 1976 and 1988 (better combustion, adaptation of gear boxes,...) and for RENAULT R5 where the complete body was changed during the same period (less aerodynamic drag,...).

PEUGEOT 104	Power (in Kw)	Fuel consumption - liters/100 km		
		90 km/h	120 km/h	Urban
1976 Model	34	6,1	8,9	8,6
1981 Model (GR)	36	5,2	6,9	7,3
1986 Model (GLS)	36	4,7	6,1	6,2

RENAULT R5	Power (in Kw)	Fuel consumption - liters/100 km		
		90 km/h	120 km/h	Urban
1976 Model	33	6,1	8,6	9
1981 Model	33	4,9	6,8	6,3
1988 Model	33	4,1	5,6	5,8

III - REAL USE AND EVOLUTION OF THE MARKET

III.1 - Real use of vehicles

In fact, we are obliged to evaluate the real use, the conditions of the using of vehicles and the impact on the real fuel consumption.

For instance, every morning, the engines are cold and one obtains a bad fuel consumption because having to heat the engine from the ambient temperature (sometimes below 0° C) to the optimal temperature (near 80 - 100° C). So, the fuel consumption is increased of 20 % in summer period to more than 40 % in winter period between cold and hot engine during the first four kilometers.

Ambient Temperature	- 10° C	0°	+ 10°	+ 20°	+ 30°
Gasoline engine	48	40	37	25	20
Diesel engine	47	43	36	23	15

Increase of fuel consumption (in %) between cold and hot engines measured during a trip of 4 kilometers after start of vehicle.

During an experimental study, we have measured the real use of vehicles with monitoring systems inside vehicles (we obtained one information per second on the engine and vehicles fonctions and one information per 10 seconds on the auxiliary equipments use (lights, wipers,...and the temperature of fluids - air, water, oil).

Some results are described below : the vehicles are used frequently for small trips and with low speeds.

	Average	weekday	differences bettween drivers
Daily uses	5.5	5.3 / 6.0	2.5 to 10.7
distances traveled km/day	38.1	30.3 / 58.8	9 to 66
trip durations min./day	57.3	52 / 76	23 to 90
consumption liters/day	3.7	3.4 / 4.9	0.8 to 6.9

Vehicles daily uses

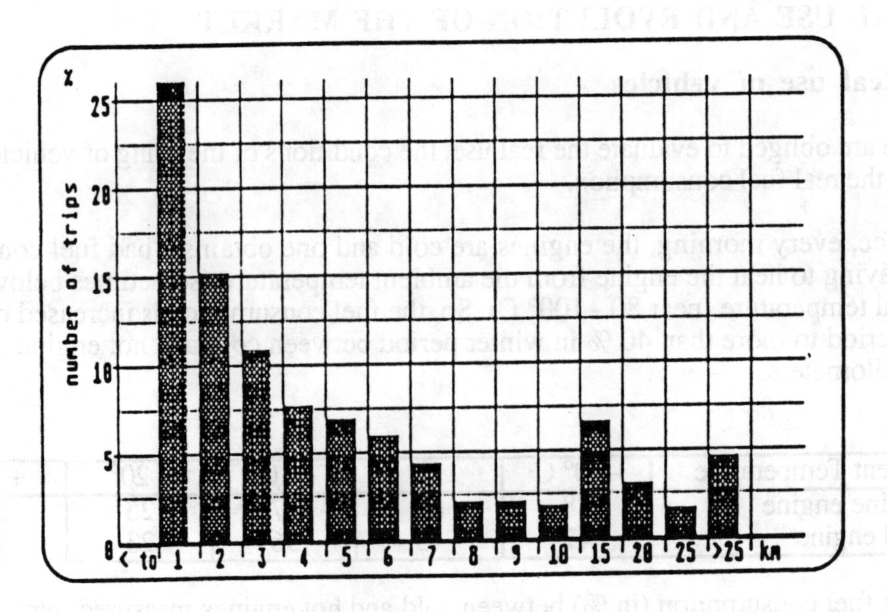

Histogram of travel lengths

III.2 - Evolution of the market - Consequences

Because of following conditions from 1986 :

- increase of economic activity
- large decrease of oil price

the evolution of the car market is going in direction of more and more speeder vehicles and in direction of higher sizes.

But if the vehicles can be driven - theoritically at high speed, they are obliged to use more power.

For instance we have indicated :

- description of small and large size vehicles
- powers to run at steady speed from 120 km/h to 220 km/h.

Model.	S (m2)	Cx	S. Cx	M (kg)
A	1,73	0,32	0,554	640
B	2,04	0,33	0,673	1290

Description of small and large vehicles

Model	120 km/h	160	180	200	220
A	17 kw	38	53	72	94
B	24 kw	52	72	96	126

Theoritical power to run at steady speed

One can see that the necessary power is

- more than three times important to run at 160 km/h than 120 km/h
- 80 % more to run at 200 km/h than to run at 160 km/h.

But the installation of such power units obliges to have a bad fuel efficiency on low speeds and specially on urban uses.

So, on steady speeds for instance with the same car body but with two different engines, the fuel consumption is higher of one to two liters from 40 km/h to 160 km/h for the speedest version.

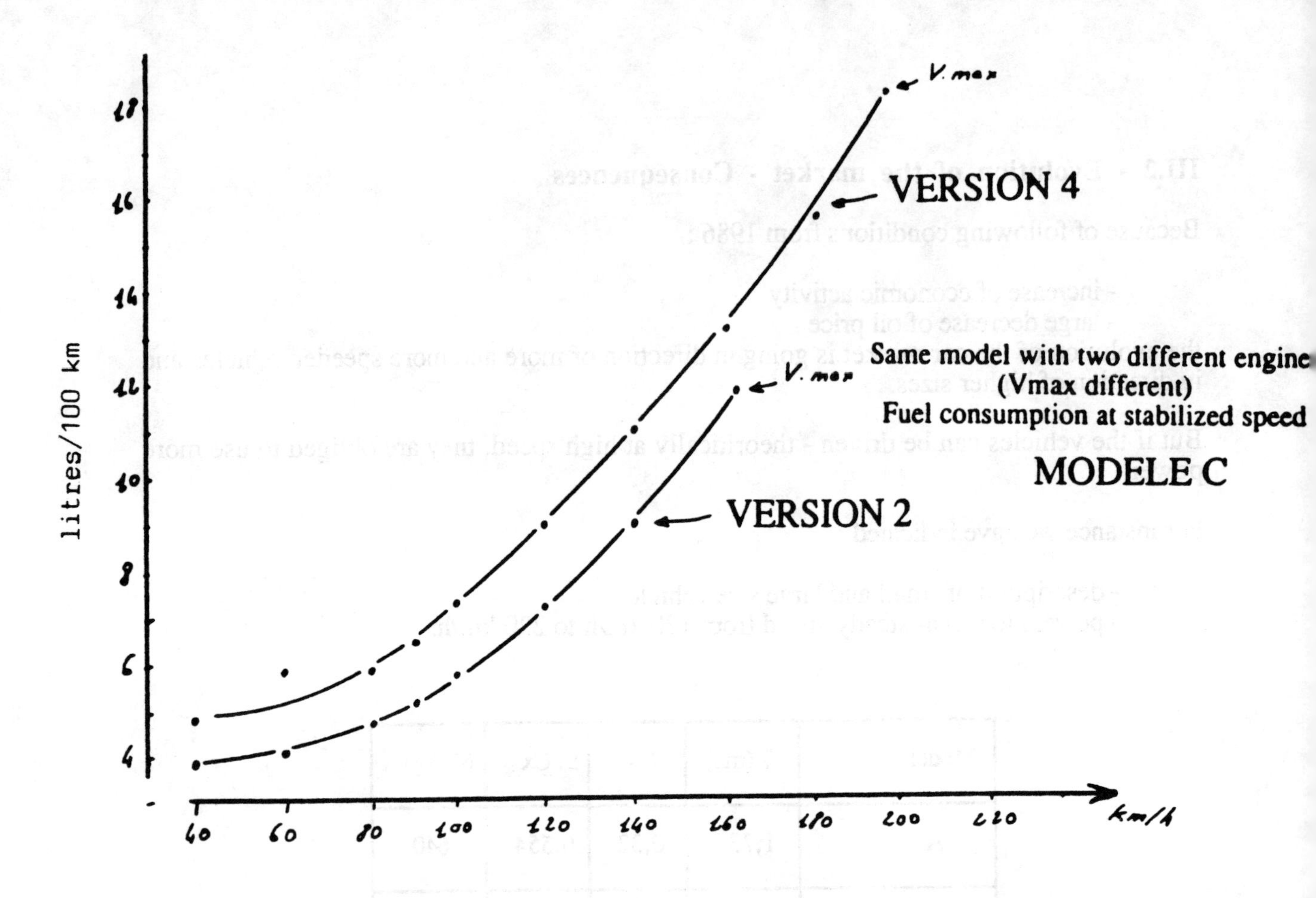

For urban use, we have selected three current models with some versions and we have indicated the fuel consumption on urban cycle : the comparisons are impressive.

Then, are indicated the urban fuel consumption (hot engine !) versus maximum speed (theoritical) for a large part of more current models sold in France in 1989.

Urban fuel consumption for three models with different engines
(two small and one medium size vehicles)

Model	Version	V max km/h	Cons. l/100 km
C	1	134	5,8
	2	162	7,2
	3	190	9,6
	4	196	9,2
D	1	155	7,8
	2	180	8,5
	3	190	9,3
	4	220	11,8
E	1	143	5,8
	2	153	7,2

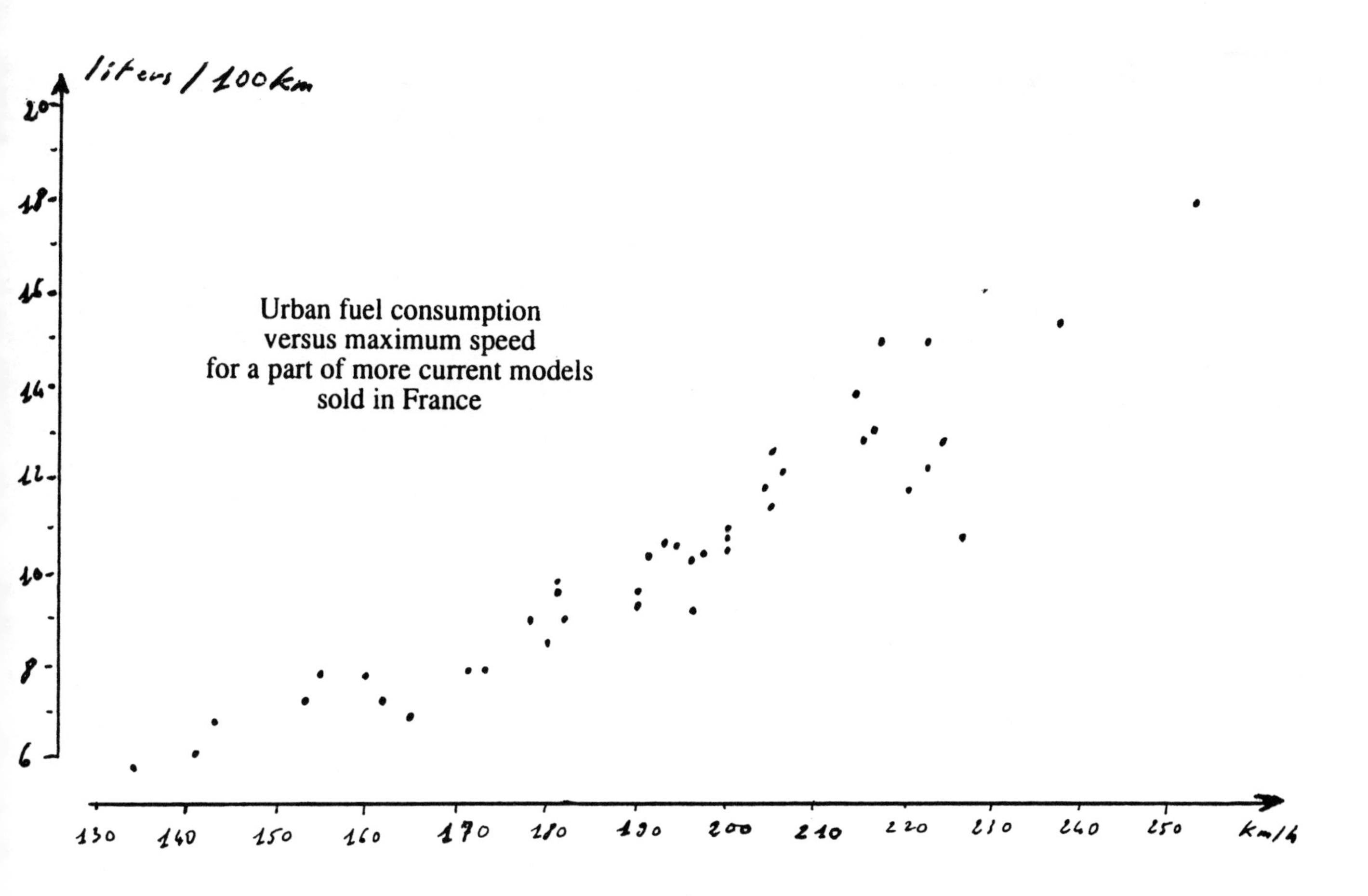

As we have said, the large part of the use of vehicles is with short trips, at low speeds and very often with cold engines.

Because of the market, more and more speeder cars are sold while the real use is with very bad fuel efficiency.

The results of demonstration programs have shown that is possible to produce small and medium size cars with a fuel consumption reduced of 20 to 40 % when we compare with commercial products.

If we accept to limit the maximum speed to 160 or 180 km/h for medium and large size vehicles (and the power) it could be possible to save near 40 to 50 % on the fuel consumption of such cars.

Voluntary limit on maximal high speed by construction and technical progress could achieve the 20 to 40 % goals of reduction for fuel consumption by cars, specially in real use.

Technology Options for Reducing Fuel Consumption and Emissions of Road Vehicles

Dipl.-Ing. Josef Brosthaus
Institute for Energy Technology
and Environmental Protection
TÜV Rheinland, D-5000 Cologne, Germany

1 Introduction

To give a comprehensive description of fuel consumption and emissions, it is absolutely essential to make a distinction between ***passenger traffic*** *and* ***freight traffic.*** *In addition to significant emission-relevant traffic parameters - passenger or freight mileage - the* ***technically feasible*** *savings potential which* ***can be realized*** *within a clearly defined period is of special and decisive importance for the development of energy consumption and climate-influencing emissions in the future. In particular, it is important to take account of relevant political boundary conditions from which trends can generally be derived on the basis of past years.*

Scenario calculations *are frequently performed to estimate future development. In this way, it is possible to combat any development errors which may become apparent and to make politicians aware of the need for action. Such a scenario on the reduction of emissions in the traffic sector by means of rational energy utilization and emission-reducing measures with prognoses for the years 2005 and 2050 was prepared for the Enquete Commission "Vorsorge zum Schutz der Erdatmosphäre" (Measures to Protect the Earth's Atmosphere) of the German Bundestag /1/. Measures and technical savings potentials discussed there are summarized in the following:*

2 *Technically feasible savings potentials*

2.1 *Measures at the* ***vehicle*** *to reduce fuel consumption*

Aerodynamic shape

c_w values for present-day compact cars: 0.32 - 0.43
Present-day medium-sized cars: 0.28 - 0.40
Large cars: 0.32 - 0.45
Achievable: 0.20 - 0.25
Fuel consumption saving: 1 to 2%
Effect on emissions: Tendency to lower them

Weight reduction

A weight reduction of 100 kg leads to a fuel saving of approx. 5%
Potential: 1 to 2%
Effect on emissions: Tendency to lower them

Reduction of rolling resistance (tyres)

Potential: 1%
Effect on emissions: Negligible

2.2 *Measures at the* ***gearbox*** *to reduce fuel consumption*

Gearbox with continuously changing transmission ratio (best point operation)

This can be realized in the medium to long term on account of ongoing development of standard automatic transmissions. Savings potential below 5% for realization after 2000.

Engine stop (automatic device to utilize flywheel momentum)

Fleet test (Golf diesel) started by VW in the police force in Lower Saxony; realization as standard equipment possible before 1995.

Saving up to 30% in dependence on the distance covered Effect on emissions: Reduction for all components

2.3 Measures at the ***engine***

2.3.1 Measures at the ***spark ignition engine*** *to reduce fuel consumption*

Reduction of friction losses

Friction losses between piston and cylinder tube

- *Mating of material ring - cylinder liner*
- *Piston ring equipment*

Improvement potential in the order of 2%

Increase in oil temperature in the part-load range, thermostatic control of oil circulation

Improvement potential: approx. 1 - 2%
Effect on emissions: Tendency to lower them

Increasing the compression ratio "ε"

Development of combustion chambers which are not susceptible to knocking. Therefore an increase of ε to between 13 and 15 is possible (Porsche TOP engine, BMFT sponsorship, $\varepsilon = 13$).

Improvement potential: approx. 10%
Effect on emissions: HC and NO_x *increase*

Combustion with high air excess

Improvements of 5% and more possible in the part-load range
Effect on emissions: Clear reduction of CO and NO_x*, HC increase*

Charging

Improvement in consumption by reducing displacement, potential approx. 10%
Effect on emissions: NO_x *increase*

Variable compression
Relatively high design expenditure, therefore not a standard feature to date
Potential: approx. 10%
Effect on emisisons: NO_x *increase*

*2.3.2 Measures at the **diesel engine** to reduce fuel consumption*

Supercharging (truck engines)

Not yet realized as a standard feature
Saving by reducing displacement
Potential: approx. 10%
Effect on emissions: NO_x *increase*

High pressure injection (truck engines)

Injection pressures exceeding 1000 bar, standard feature (KHD), development partly within the scope of the BMFT project "Low Noise Diesel Engines". Improvement of fuel consumption in the order of 5 to 10%.

Effect on emissions: Reduction of NO_x and particles, reduction of combustion noise.

Electronic control of injection behaviour in dependence on load and speed

Effect comparable to the use of map ignition in the spark ignition engine
Already standard equipment in BMW
Improvement of fuel consumption more substantial for the direct injection engine than the chamber engine
Potential: 5 - 10%
Effect on emissions: Reduction for all components

Direct injection (car engines)

Problems: Loud combustion noise, substantial increase in NO_x emissions, to date the best results were obtained with the Elsbett-Duotherm diesel engine (BMFT project), US limits were met.
Fuel consumption advantage over swirl chamber engine: more than 10%

3 Examples of research vehicles with favourable fuel consumption

Some automobile manufacturers have presented new solutions for future automobile construction in so-called ***research cars*** *or* ***concept studies****. Those cars which are most up-to-date and most interesting with regard to low energy consumption are the VW Öko-Polo, the Renault Vesta 2 and the VW Futura.*

The ***Öko-Polo*** *constitues optimization of the drive of a small and economic car. The charged 2-cylinder diesel engine with direct injection is equipped with an automatic device to utilize fly-*

wheel momentum, which declutches and switches off the engine during coasting or when the car comes to a standstill. The flywheel is not used to restart the engine. NO_x *emissions are reduced with charge air cooling and exhaust gas recirculation. As expected, CO and HC emissions are clearly below the US limits of the US-FTP-75-cycle, while* NO_x *and particle emissions (without particle filter) are only slightly below them. From autumn 1989, 50 vehicles will be tested during day-to-day operation in Berlin. They will be equipped with particle filters which will probably increase fuel consumption. On average, the consumption is around 3 - 4 l/100 km. It would be much lower if the body weight and aerodynamic drag were lower.*

Development work on the Renault VESTA 2 did not center around optimization of the drive but rather the bodywork. By using low-weight materials, reducing the weight of existing parts and keeping interior fittings relatively simple, it was possible to construct a vehicle with a c_w *value of 0.19 which is around 40% lighter than the Polo - although it has almost the same outside dimensions. This means that its fuel consumption is much lower than that of the Polo at 90 and 120 km/h and that its maximum speed is identical to that of the Polo although the rated power is one third lower. A small, low-power spark ignition engine was selected to bypass unfavourable part-load consumption.*

The fuel consumption of the VESTA 2 is around 3 l/100 km during mixed operation, and is therefore slightly better than that of the Öko-Polo. Both vehicles have performed record test drives with a fuel consumption of 1.9 l/100 km (VESTA 2) and 1.7 l/100 km respectively. However, the VESTA 2 is not planned as a standard production vehicle in this form.

VW presented the research car IRVW-Futura at the IAA 1989. Its engine concept is particularly interesting and the technical data have now been published /2/. The aim of engine development was to obtain a powerful and comfortable spark ignition engine, which met the US limits, had very low consumption and low noise. The main features of the engine are permanent lean-burn adjustment, direct injection, charging with charge air cooling, evaporation cooling and an oxidation catalytic converter. Most significant fuel savings were achieved during part-load operation. A Golf which was equipped with this engine required 6.7 l gasoline/ 100 km in the ECE city cycle, the less powerful Golf Turbo-Diesel used 6.3 l diesel/100 km.

4 Unsolved problems and necessary analytical and research activities

4.1 Motorised passenger traffic

Some of the fuel-saving and emission-reducing technical measures in the "traffic" sector can lead to an increase in other emissions, in some cases emissions which also exert an influence on climate. Therefore verified data on the following subjects should be made available as soon as possible:

- *The extent to which NO_x, HC and CO emissions, which do not exert a direct influence on climate, contribute to ozone formation and other greenhouse processes. In this case, it is important to consider the different effect of the emissions in dependence on the location at which they originated.*

- *The climatic importance of other air pollutants (water input by aircraft, sulphur oxides, (carbon) particles and dust).*

Furthermore, estimates of methane emissions are based on a few, non-representative data. As far as we know, detailed measurements, in particular of car exhaust emissions have already been organized.

*In **car traffic**, problems relating to the representative determination of: mileage, driving behaviour, emission factors of the cars including the numerous pollutant-reducing systems (long-term stability) have been known for a long time and some of them are the subject of projects being carried out in the Federal Republic of Germany. To date, however, the balance of energy and emissions for the production, maintenance and waste disposal of the transport system "car" has not been considered in depth. Some measures for the reduction of climate-influencing emissions could affect this balance, leading to both positive and negative feedback.*

*4.2 **Motorised freight traffic***

The data base for freight traffic is scanty. Data are frequently not available in the form required for validated energy and emission-related statements. For example, statistics are often orientated towards aims which are of importance from the entrepreneur's viewpoint (profitability aspects etc.) but which are not always suitable as a basis for energy and emission statements.

Furthermore, for example, passenger mileage in tkm as a standard reference variable is linked with uncertainties which cannot be exactly assessed. This is because it is calculated as a secondary variable from the traffic volume using the transport distance and degree of loading, parameters for which available information is frequently incomplete.

With reference to freight traffic, a uniform registration system should be used to record traffic data. Its purpose should not only be to collect passenger mileage data but also to allow statements to be made on the driving behaviour of the respective means of transportation. For example, information on accelerating and decelerating processes during traction is of interest since their influence on the energy and emission level is not to be underestimated.

On the emission side, more representative studies of emissions from vehicles on the road would need to be performed. Such data will be available for road freight traffic on completion of a research project commissioned by the Federal Environmental Agency (Umweltbundesamt) to determine emission factors for the commercial vehicle sector which is currently underway in the Federal Republic of Germany /3/.

In the long term, it would be desirable to carry out research projects on the possible utilization of alternative fuels for freight traffic and their effects on the environment.

5 References

/1/ *TÜV, IFEU:*
Emissionsminderung durch rationelle Energienutzung und emissionsmindernde Maßnahmen im Verkehrssektor
Report for the Enquete Commission "Vorsorge zum Schutz der Erdatmosphäre" of the German Bundestag
TÜV Rheinland, Cologne; IFEU, Heidelberg
October 1989

/2/ *Emmenthal, K.-D. et al:*
Motor mit Benzin-Direkteinspritzung und Verdampfungskühlung für das VW-Forschungsauto IRVW-Futura
Motortechnische Zeitschrift 50 (1989) Vol. 9
Franckh-Kosmos-Verlag, Stuttgart

/3/ *R+D project*
Das Abgas-Emissionsverhalten von Nutzfahrzeugen in der Bundesrepublik Deutschland im Bezugsjahr 1990
TÜV Rheinland, Cologne and RWTÜV, Essen for the Federal Environmental Agency, Berlin
(under preparation)

THE INTERACTION BETWEEN EXHAUST EMISSIONS AND FUEL ECONOMY FOR PASSENGER CAR ENGINES

C.C.J. French, C.H. Such

Ricardo Consulting Engineers plc

Conventional Engines and Fuels

The diesel engine offers the best fuel efficiency for almost all land transport applications, apart from very high speed operation of passenger cars and two-wheeled vehicles.

Currently, with the exception of the Fiat Croma and Austin Rover Montego direct injection [DI] diesels, all passenger car diesel engines use combustion systems with indirect injection [IDI], mostly of the Ricardo Comet V swirl chamber type or of the Daimler-Benz prechamber type. Compared with the DI engines, the IDI systems offer superior exhaust emissions and noise whereas the DI has better fuel economy.

Diesel cars with IDI engines have a volumetric fuel economy advantage of 20%-25% in mixed driving [Euromix] of approximately equal performance.

The 20%-25% economy advantage still holds good if both engine types are modified to meet strict emission regulations such as the 1987 US Federal limits which will apply in some non-EEC countries.

In practice, the economy advantage of the diesel is greater than the Euromix data suggest as these do not reflect the advantage of the diesel during engine warm-up.

Cars with DI engines have a potential economy advantage of up to 20% compared with IDI cars but, in practice, taking into account the future emissions limits in Europe, this is likely to be nearer 10%-15%. The trade-off between fuel consumption and exhaust emissions for gasoline and diesel vehicles is shown in Figure 1. At one time, lean burn gasoline engines threatened the advantage of the diesel, but, in the absence of a catalyst which can reduce NOx at lean mixtures, the lean burn engine seems to be unable to meet future EC limits.

Regarding CO_2 emissions, even taking into account the greater density of diesel fuel, the diesel offers a CO_2 advantage of some 10%-15% for IDI cars and 20%-25% for DI cars compared with gasoline cars fitted with three way catalysts. Apart from the steady improvements in conventional gasoline engines, this differential is unlikely to change substantially unless direct injected, stratified charge [DISC] gasoline engines can be developed to meet future emissions limits.

FUEL ECONOMY/EXHAUST EMISSIONS FOR MID-CLASS VEHICLE

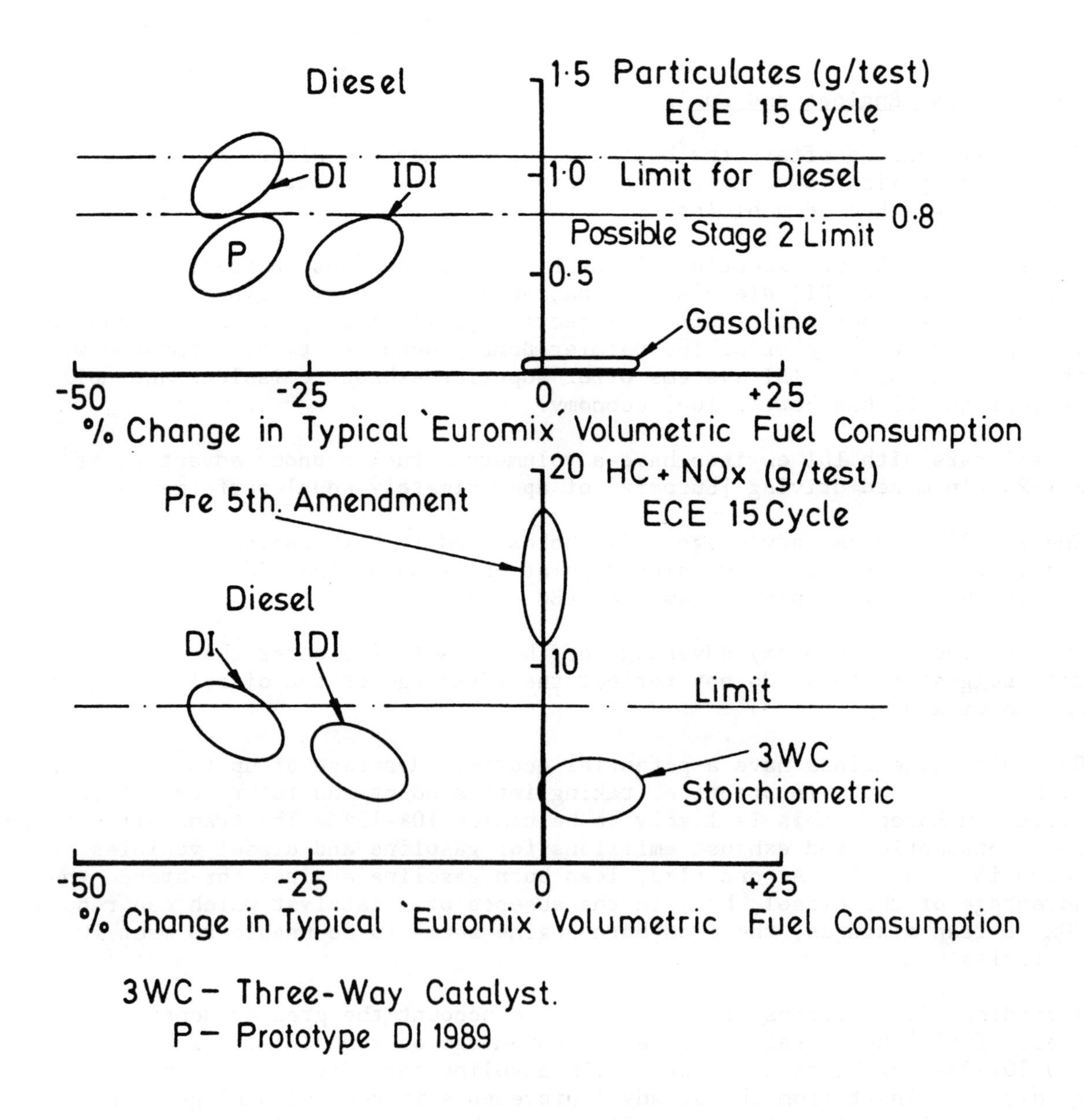

Figure 1

The big question marks which hang over the diesel are its NOx and particulate emissions. There is no doubt that, due to its lean burn combustion, the diesel can not achieve the very low NOx levels which the gasoline engine with three way catalyst is capable of. There is research on catalytic and chemical methods of reducing NOx from diesel engines but these are still in the laboratory stage and practical and cost effective solutions are a long way off.

Regarding particulates, recent improvements in combustion development, fuel injection equipment [including electronic control] and the use of oxidation catalysts have all helped to reduce total particulates, and the catalyst may reduce the toxic fraction of the particulate. Further improvements are likely from these directions and also from the diesel fuel itself: especially a reduction in sulphur content will help to reduce particulate and make oxidation catalysts more feasible.

Overall, it seems that the diesel is still able to provide a benefit in terms of low CO_2 emissions in the European context and therefore emissions limits should not be set so tight as to rule out the diesel.

Alternative Fuels

It is unfortunate that all alternative fuels have some disadvantage as compared with today's petroleum based fuels, not least of which is increased cost. This is not to say, however, that they may not also have some advantages.

Gaseous fuels give problems of vehicle range due to the practical limits of storage container size. Natural gas is of course readily available in many parts of the world. Due to the higher H/C ratio of methane compared with gasoline as a fuel, its use would reduce CO_2 emissions but methane itself is a greenhouse gas.

Conversion of natural gas to methane is possible, although the production process is likely to be expensive and losses in the process reduce the advantage in overall CO_2 emissions as compared with the use of straight natural gas.

Due to their low volumetric energy density, the use of alcohol fuels roughly halves the vehicle range unless a larger fuel tank can be fitted. Alcohol fuels also give cold start problems due to the much higher heats of vaporization. A further point is that, in view of the uncertain nature of the supply of alcohol fuels, vehicles would need to be dual-fuel, i.e. able to run on both gasoline and alcohol, which increases the cost and complexity considerably.

Oxygenates such as methyl tertiary butyl ether [MTBE] will continue to find a role in improving octane numbers in spark-ignited engines.

The use of alternative fuels in diesel engines poses very severe problems. The most practical solution would be to convert to spark ignition but experimental work continues on the use of ignition improvers or alternatively on the use of exhaust gas recirculation, glow plugs and a high compression ratio.

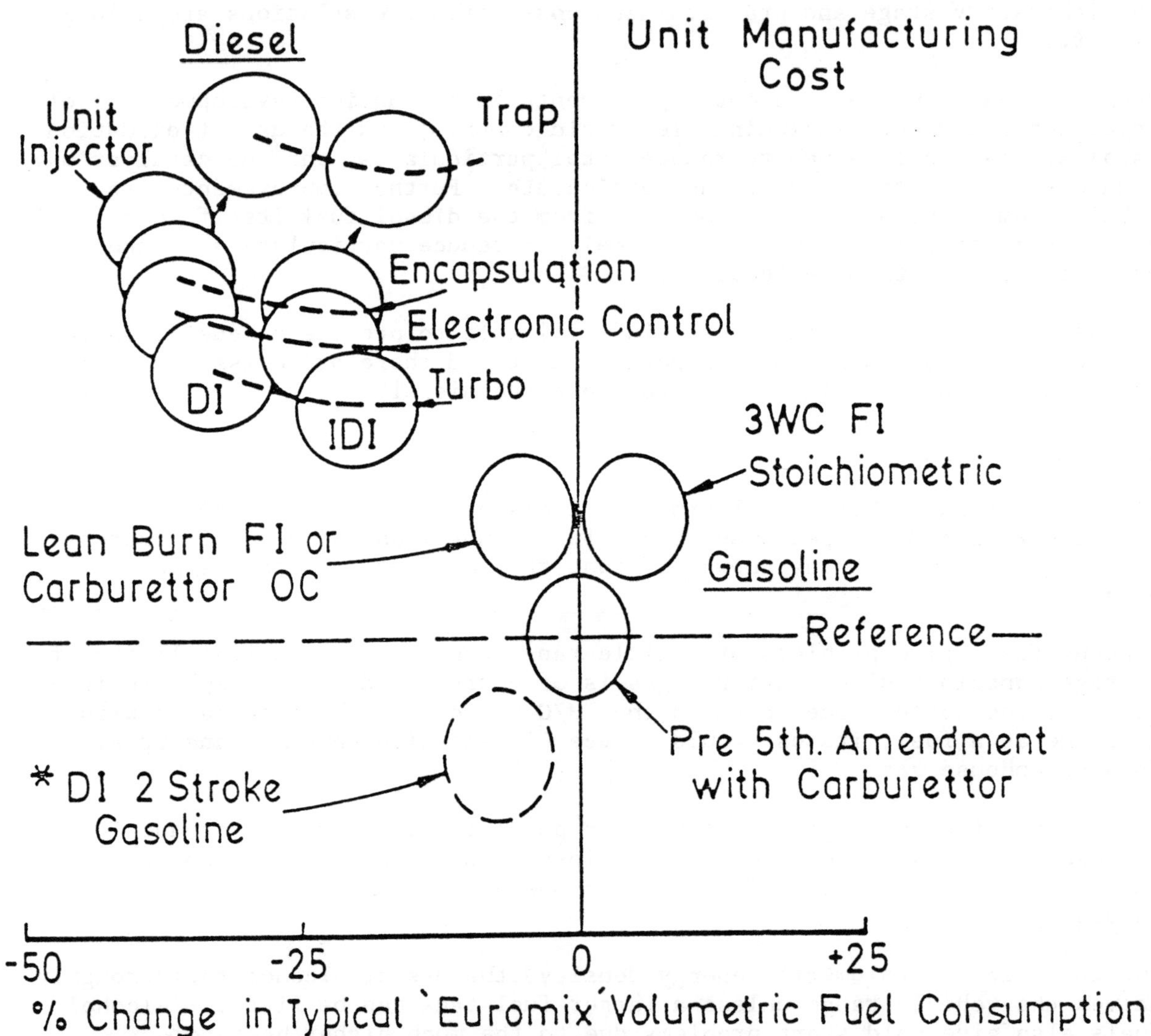

3WC – Three-Way Catalyst.
OC – Oxidation Catalyst.
FI – Fuel Injection.

Fuel Economy / Manufacturing Cost for Various Versions of a Basic Engine All with the Same Power, Calibrated to EEC Emissions Limits (5th. Amendment) All Engines Made on the Same Production Line Except 2-Stroke.

Alternative Engines are unlikely to be employed in anything but a long time scale.

Stirling engines tend to be heavy and bulky and while their efficiency has improved and can now approach that of a spark ignition engine, problems remain in retaining the charge of hydrogen or helium in the engine.

Gas turbine efficiency approaching that of the spark ignition engine, let alone that of the diesel, requires much higher operating temperatures than are currently usable. Engineering ceramics may one day make that possible but they are currently not available.

The fuel cell requires a major breakthrough in bulk, cost and weight although it is potentially of interest in the long term since, as it is not a heat engine, it does not have a Carnot limit of efficiency.

Hybrid vehicles with storage devices such as flywheels or batteries offer possibilities of reduced fuel consumption and hence reduce exhaust emissions but at the expense of complexity and increased cost.

The all-electric vehicle requires a breakthrough in battery technology to achieve an acceptable range, performance and overall vehicle weight. The electricity must also be generated and in many areas of the world where nuclear energy appears to be unacceptable this will involve the burning of fossil fuels and hence, particularly when transmission losses are taken into account, the overall production of exhaust emissions may well be increased.

SUMMARY

Of the conventional technologies, diesel engines offer the greatest potential for CO_2 reduction but problems of NOx and particulates need to be solved.

Gasoline engines with three-way catalysts provide the lowest emissions of hydrocarbons, CO and NOx. Research on improved fuel economy and hence lower CO_2 is continuing.

Significant reductions in CO_2 may be realized by engines running on gaseous fuels but there are practical problems related to vehicle range. Depending on the production process, methanol may reduce CO_2 emissions by a small amount.

Although much effort has gone into the development of alternative power plants for vehicles, we have yet to see practical solutions suitable for mass production.

Based on the existing technology and the expected fuel economy improvements in the future, the goal of achieving a 20%-30% reduction in CO_2 will not be solved by changes to engines or transmissions alone. If the goal is to be achieved, then government regulation or fiscal incentive related to vehicle usage would seem essential.

FUEL ECONOMY IMPROVEMENT OF PASSENGER CARS

Leopoldo Chinaglia

Consiglio Scientifico - Istituto Motori CNR - Napoli

1. Research Car

In order to review the options to improve the fuel economy of passenger cars, the results of the "Research Car Project", which the Consiglio Nazionale delle Ricerche - Progetto Finalizzato Trasporti has performed (the main contractor was the Centro Ricerche Fiat), are discussed.

The project yielded prototypes exhibiting an outstanding fuel economy. The measured fuel consumption of three prototypes, having alternative engines, is shown in table 1.

Table 1

RESEARCH CAR FUEL CONSUMPTION RESULTS (l/100 Km)

	ECE 15	90 Km/h
PROTOTYPE 1	5.2	3.7
PROTOTYPE 2	3.4	2.7
PROTOTYPE 3	3	2.5

The features of the propulsion plants of the above prototypes are shown in table 2.

Table 2

PROPULSION PLANT FEATURES

PROTOTYPE 1	. SPARK IGNITION, HIGH COMPRESSION RATIO, LEAN BURN, 2 CYLINDER ENGINE MANUAL 5 SPEED TRANSMISSION
PROTOTYPE 2	. DIRECT INJECTION, TURBOCHARGED DIESEL ENGINE . CONTINUOSLY VARIABLE TRANSMISSION (RUBBER BELT, ELECTRONIC CONTROL) . START/STOP OPERATION
PROTOTYPE 3	. SAME AS PROTOTYPE 2 BUT WITH IMPROVED TRANSMISSION CONTROL. (MODULATION OF SHEAVES FORCE AS A FUNCTION OF LOAD AND RATIO)

The research car had a carrying capacity of 5 occupants (95 percentile) in comfort position.

Performances and main parameters affecting fuel economy and are shown in table 3.

Table 3

RESEARCH CAR PARAMETERS AND PERFORMANCE RESULTS

MASS	540 Kg (SPARK IGNITION) 590 Kg (DIESEL)	
S.Cx	0.45 m^2	Cx = 0.25
ROLLING RESISTANCE	8 Kg/t	
SPEED	135 Km/h	
ACCELERATION 0-400 m	23 s	
ACCELERATION 0-1000 m	46 s	

Comfort, handling, durability of the research car were equal to or better than those of the production car which was taken as reference (representative of the lower segment in Europe in the early 80's).

For comparison the fuel consumption of the reference car is shown in table 4, while table 5 shows parameters and performance data.

Table 4

REFERENCE CAR FUEL CONSUMPTION (l/100 Km)

ECE 15	90 Km/h
7.5	5.5

Table 5

REFERENCE CAR PARAMETERS AND PERFORMANCE DATA

MASS	650 Kg	
S.Cx	0.7 m^2	Cx=0.4
ROLLING RESISTANCE	12.5 Kg/t	
S.I. CARBURETTED, CONVENTIONAL IGNITION ENGINE		
MANUAL 4 SPEED TRANSMISSION		
SPEED	125 Km/h	
ACCELERATION 0-400 m	21.2 s	
ACCELERATION 0-1000 m	41.2 s	

The improvement of fuel efficiency of the research car with respect to the reference car is of the order of 30% for the spark ignition alternative and more than 50% for the direct injection Diesel alternative.

The role of the single car parameters in this improvement is shown in table 6.

This table presents the effect of reductions in mass, aerodynamic penetration and rolling resistance. The two latter parameters have played an important role in reducing the 90 Km/h fuel consumption.

Table 6

PERCENT DECREASE OF FUEL CONSUMPTION FOR EACH PERCENT DECREASE OF THE INDICATED PARAMETER

	ECE 15	90 Km/h
VEHICLE MASS	0.65%	0.4 %
AEROD. RESISTANCE	0.18%	0.5 %
ROLLING RESISTANCE	0.15%	0.25%

The benefit of the Continuosly Variable Transmission is of the order of 6% in the Urban Cycle and 4% at 90 Km/h, provided that a very sophisticated control strategy is used. Start-Stop operation contributes 10%, in urban driving.

The engine plays a major role on fuel economy and the direct injection diesel is the single most effective measure to improve fuel economy.

This remark introduces the discussion on the emissions fuel-economy trade off of the next section.

2. Near term options for fuel economy improvement in Europe

The emission standards which will become effective in the EEC in 1992 are essentially equivalent to the Federal U.S.A. standards

The effect will be the general use, on new spark ignition models, of electronic control of fuel and ignition systems, as a consequence of the three way cathalist requirement.

The small decrease in fuel economy, due to stoichiometric operation, will be probably offset by less dispersion, resulting in "in use" fuel economy gain. To this effect are important improved electronic control systems, including engine diagnostic and optimal transient control strategies.

The prechamber diesel, which has become "European Technology" after its disappearance in the U.S., can meet the 1992 EEC standards. The capability of the direct injection Diesel to meet the EEC 92 standards is still a development theme. As it was mentioned before, the diffusion of this type of engine is desirable to improve the fuel economy.

In table 7 a spark ignition engine, a prechamber diesel and a direct injection diesel are compared in terms of NOx, Particulate and CO_2 emissions (the latter being indicative of fuel usage).

The results refer to the same car (3000 lb inertia mass) tested on the Urban FTP 75 Cycle, in the following three versions:

- 2 liter Spark Ignition Engine, Electronic Fuel Injection, 3way Cathalist
- 2.5 liter Prechamber Diesel Turbocharged, Exhaust Gas Recirculation Electronically Controlled
- 2 liter Direct Injection Diesel, Turbocharged, Fuel Injection Pump Electronic Control, Closed Loop E.G.R., Trap Oxidizer.

Table 7

COMPARISON OF THREE "ECOLOGICAL" ENGINES

(NOx AND PARTICULATE VALUES ARE NORMALIZED TO U.S. LIMITS - CO_2 IS NORMALIZED TO SPARK IGNITION)

ENGINE TYPE	CO_2	NOx	PARTIC.
SPARK IGNITION	1.0	0.2	0.2
PRECHAMBER DIESEL	0.9	0.8	0.7
DIRECT INJECTION DIESEL	0.8	0.9	0.9

All the above configurations meet the EEC 92 Standards. However the diesels, in particular the direct injection, require electronic fuel system control and/or particulate filters, which are essentially unproven technology. In addition fuel quality improvements (sulfur and aromatics content) are essential.

European Manufactures have shown a strong committement to direct injection development. Uncertain emission scenarios might discourage the maintenance of an high level of effort. In the USA a further reduction of NOx limits is being discussed; pressures in the EEC to follow the example would impair the survival of the diesel for passenger cars.

At the engine level, further near term options of improvement are limited.

The lean burn engine, which exhibits a fuel economy improvement of 10-15% with the respect to stoichiometric operation, will not have large diffusion due to the EEC 92 standards.
This concept was in fact developed for less stringent requirements.

Spark ignition direct injection stratified charge engines, including two stroke engines, will also probably remain a research theme. These engine types are suitable for ultralean operation.

In Italy passenger cars fueled with compressed natural gas and L.P.G. are already diffused. In terms of CO_2, compressed natural gas has margins "vis a vis" of diesel fuel. The original gasoline engines are transformed by specialists and configurations optimizing the emissions are under development.

High power density engines (multivalves, suction control) allow mass reduction and for this reason offer some fuel economy positive effects.

At the vehicle level, the improvement in aerodynamic penetration which was realized in the 80's leaves limited room for further reduction. Typically the Cx value of recent models is 0.30; the next step could be 0.22-0.25, contributing several percent to fuel economy.

The reduction in car mass has a remaining potential to be exploited. The progress in this respect has been limited during the past decade, because the addition of equipment tended to offset the gains in structure.

In conclusions, the most effective way of improving fuel economy in cars in the near future is the diffusion of the direct injection diesel. This requires the use of emission control technologies wich are still developmental.

These technologies are compatible with 92 EEC standards but unlikely to cope with further reduction of NOx limits.

The Need of Low Consuming and Emitting Automobiles

Norbert Gorißen
Umweltbundesamt, Berlin, West Germany

1. Fuel Consumption and Exhaust Gas Emissions of Automobiles in the Federal Republic of Germany 1966 and 1986

In Table 1 the data of fuel consumption and exhaust gas emissions for passenger cars, lorries and busses 1966 and 1986 are given. Additionally there are indicated the numbers of vehicles and the mileage. The given data underline the growing importance of fuel consumption and exhaust gas emissions of motor vehicles.

The fuel consumption of passenger cars grew between 1966 and 1986 from 70 g fuel/km to 80 g fuel/km, for lorries and busses it grew from 195 g fuel/km to 225 g fuel/km.

Table 1:

[in 1000 tons] 1966	fuel consumption	CO_2	CO	HC	NO_x	particles	number of vehicles [in 1000]	mileage [in billion km]
passenger cars	11.209	35.500	6.223	568	345	10	10.302	159,3
lorries and busses	5.183	11.300	68	55	265	19	2.293	26,6
1986								
passenger cars	26.846 +140%	84.300 +137%	6.106 -2%	926 +64%	1.057 +206%	20 +100%	26.917 +162%	336,2 +111%
lorries and busses	9.486 +83%	20.700 +83%	129 +90%	103 +87%	502 +89%	37 +95%	3.407 +49%	42,2 +59%

2. Influence factors on fuel consumption

The total fuel consumption of a vehicle fleet is the product of:

- number of vehicles,
- mileage per vehicle,
- specific fuel consumption per km driven,
- correction factor for the individual driving behaviour.

The specific fuel consumption values for the total fleet passenger cars and for the yearly new registered passenger cars are given in table 2. These values are evaluated by using realistic test-procedures, instead of the DIN-test-procedure, which are always considerably lower. It is obviously, that there is only a slow decrease in specific fuel consumption between 1978 and 1988. That is due to two reasons:

- the registered car fleet is getting older,
- more and more cars with larger swept volume and more engine power are getting registered.

Table 2: Specific fuel consumption values, as measured in tests

[fuel/100 km]	1978	1979	1980	1981	1982	1983	1984	1985	1986	1987	1988
total fleet of passenger cars	10,9	10,79	10,85	10,75	10,69	10,68	10,64	10,59	10,5	10,36	10,38
total fleet of Otto passenger cars	10,98	10,84	10,95	10,89	10,9	10,91	10,9	10,9	10,9	10,8	10,66
total fleet of Diesel passenger cars	9,5	9,58	9,69	9,4	9,1	9,1	9,0	8,8	8,5	8,3	8,29
new registered Otto passenger cars	11,5	11,4	11,1	11,0	10,7	10,3	10,0	10,0	9,8	9,7	*
new registered Diesel passenger cars	9,4	9,5	9,4	8,4	8,6	8,2	7,8	7,7	7,7	7,9	*

* no yet available

In figure 1 the development of the specific fuel consumption of all registered passenger cars in the FRG is shown.

FIG. 1: mean specific fuel consumption of the total fleet of passenger cars
(in l fuel/100 km)

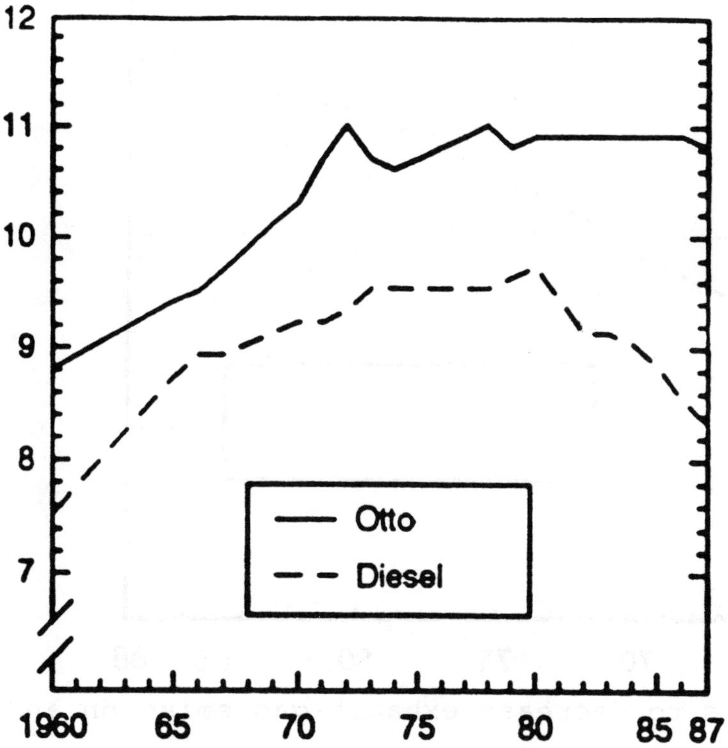

Figure 2 shows the development of the swept volume classes, figure 3 illustrates the steady increase of engine power of the fleet of all registered vehicles in the Federal Republic of Germany.

Fig. 2: Reg. Passenger-Cars in different Swept-Volume classes

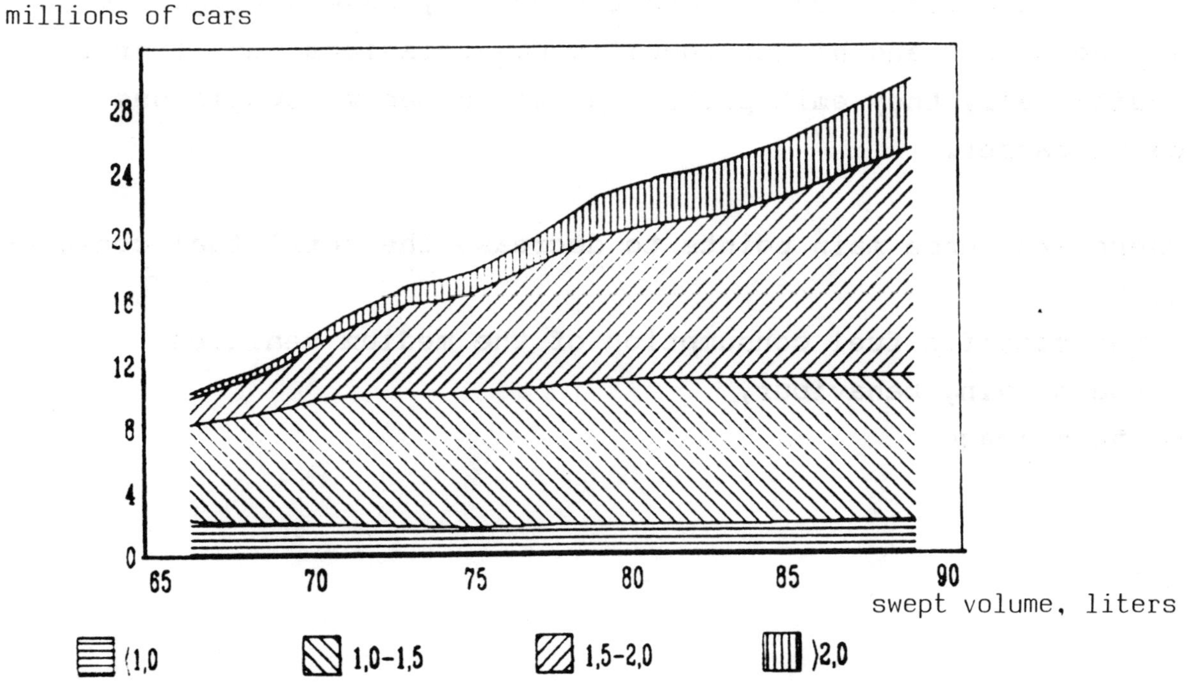

FIG. 3: mean engine power of the total fleet of passenger cars (in kW)

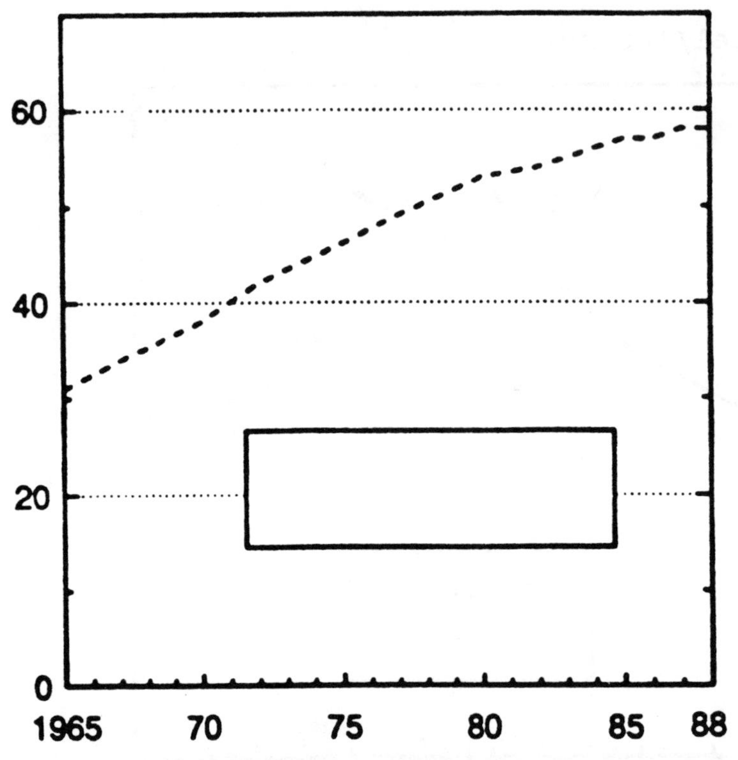

3. Starting points to decrease exhaust gas emission and fuel consumption of motor vehicles

The most effective way to reduce exhaust gas emission of passenger cars, is the implementation of closed loop three way catalysts. Diesel engines may have less volumetric fuel consumption than Otto engines. On the other hand they have a bigger amount of NO_x emissions and higher specific CO_2-emissions per liter fuel. Additionally they emit particles, which become suspicions to cause cancer.

There are three main points to decrease the total fuel consumption:

- the specific fuel consumption of the medium vehicles,
- the driving behaviour,
- the mileage.

3.1 Specific fuel consumption

The most important influence factors of the specific fuel consumption are the vehicle mass and the engine power of the vehicle.

Figure 4 illustrates the dependency of specific fuel consumption and vehicle mass (0,7 l/100 km per 100 kg vehicle mass), figure 5 the dependency of specific fuel consumption and engine power (0,45 l/100 km per 10 kW engine power) evaluated by IFEU Heidelberg, using fuel consumption values measured by the ADAC. These dependences are valid for new often sold vehicles in the Federal Republic of Germany in the year 1989. So it is obvious, that by reducing the engine power and/or the vehicle mass also the specific fuel consumption can be reduced.

FIG. 4: mean specific fuel consumption (in l fuel/100 km) in dependence of vehicle mass (in kg)

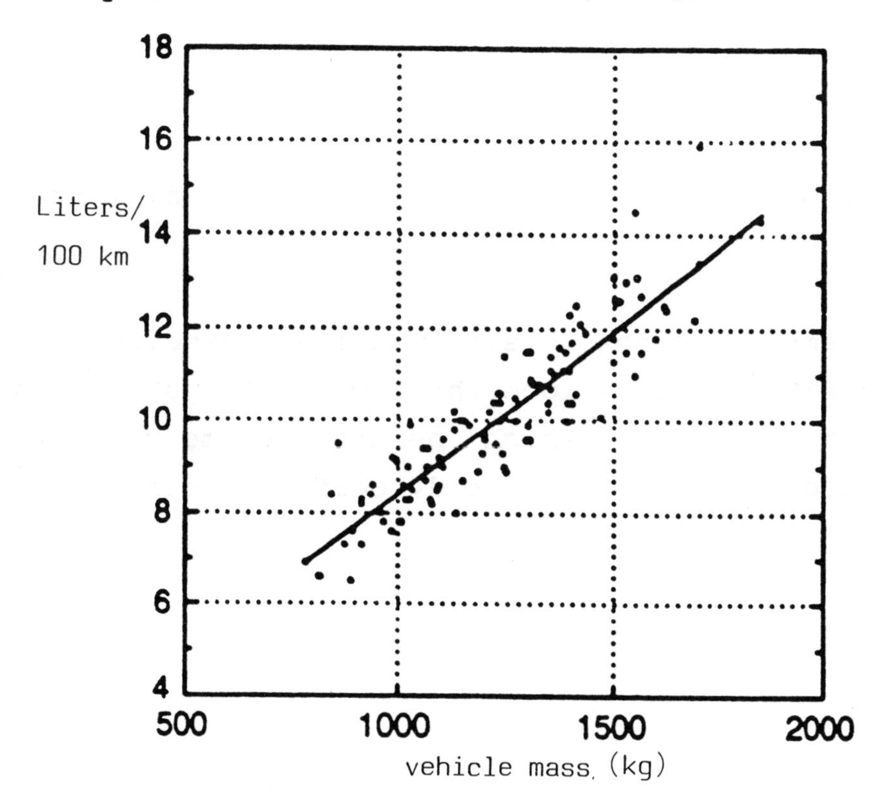

FIG. 5: mean specific fuel consumption (in l fuel/100 km) in dependence of engine power (in kW)

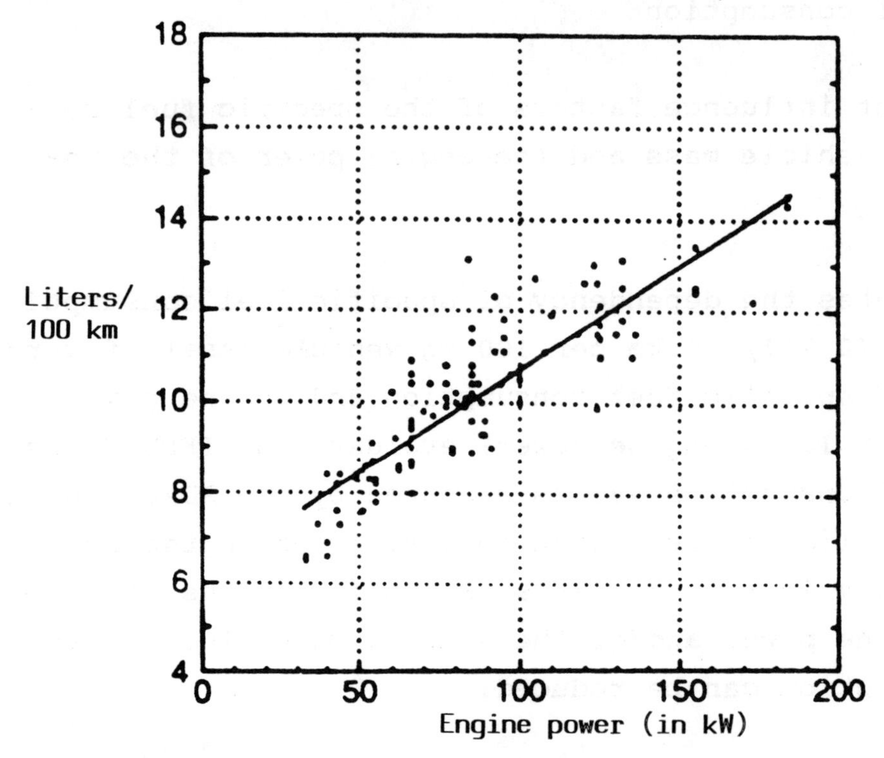

Other technical aspects to influence the specific fuel consumption are:

- air resistance,
- rolling friction (e.g. tire width),
- laying out of the gearing,
- other motor parameters (compression, motor friction, mixture control, etc.).

Summarized, it seems to be possible to realize a gasoline passenger car equipped with a closed loop three-way-catalyst, using already well known techniques, which doesn't consume more than 5 l/100 km in average and which is already comfortable for the user.

3.2 Driving behaviour

The next important aspect of reducing fuel consumption is the driving behaviour, especially the driven speed. As evaluated at 15 measuring points by the BaSt the driven speed of passenger cars is rising (see figure 6). Depending on the surveillance and on the level of a possible speed limit the total fuel consumption of all passenger cars can be reduced between 1,5% (speed limit 120 km/h on highways) and 8,5% (speed limits 100 km/h on highways, 80 km/h on rural roads, 30 km/h on residential streets in towns, sharp surveillance).

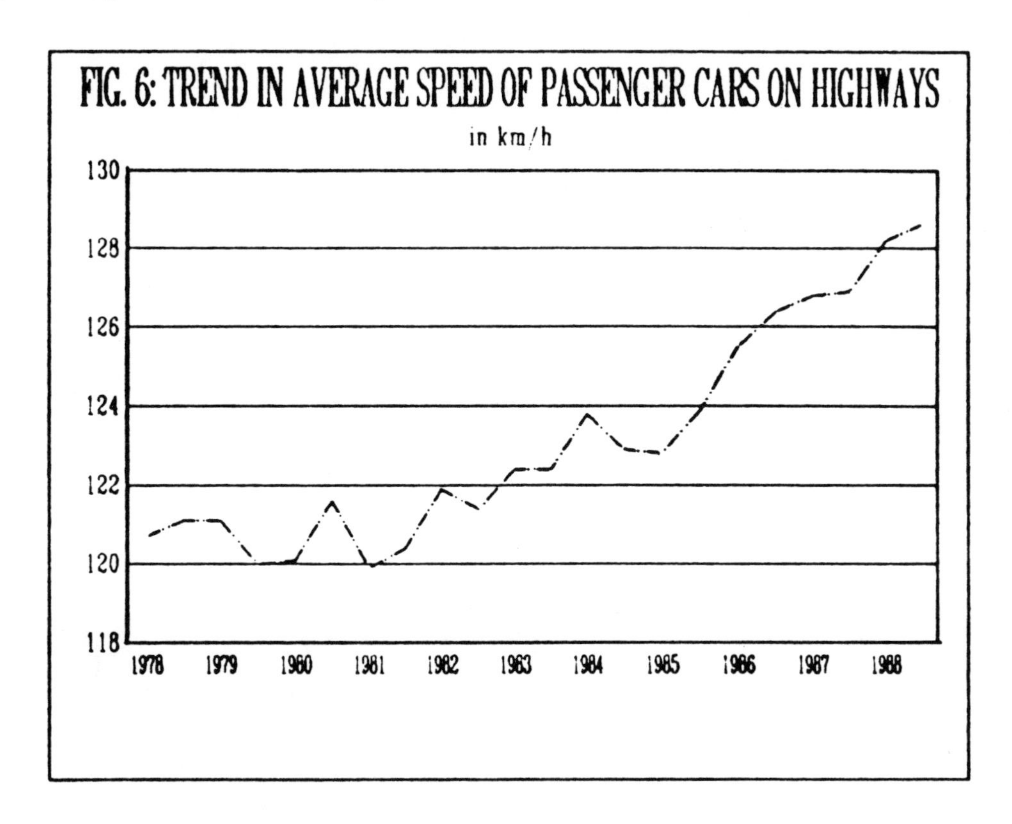

3.3 Mileage

The most effective way to reduce fuel consumption is to minimize the mileage. This target can be met by a number measures, like:

- increase of variable costs (e.g. fuel tax)
- restrictions in car use (e.g. parking management, road pricing)
- improvements in public transport (e.g. prices, infrastructure).

DOWNSIZING OF AUTOMOTIVE SPARK IGNITION ENGINES

J F Bingham
IC Engines Division
National Engineering Laboratory
East Kilbride
Glasgow

S U M M A R Y

Downsizing of engines for passenger cars would result in improvements in fuel economy but at the expense of performance. This could be offset by intelligent supercharging under the control of the engine management system. Despite this it appears that consumer education would be required before downsized engines would gain widespread acceptance.

1 INTRODUCTION

Proposals to reduce fuel consumption of road vehicles must be assessed on a total energy basis so that any saving in the quantity of fuel used by vehicles is compared firstly with the energy expended to introduce and maintain the new technology and secondly with any likely increase in energy consumers in other transport sectors. A major obstacle to the introduction of low consumption passenger cars is the expectation by customers of the high levels of performance which have become a leading feature of many automobile marketing strategies. The introduction of new technology to reduce fuel consumption should therefore be allied to a government programme of consumers education as one element of a national or international total energy policy.

2 DOWNSIZING OF ENGINES

Reducing the swept volume of reciprocating engines produces a number of benefits but also creates some problems. The major drawback is that of reduced levels of performance which can be overcome to some extent by adopting intelligent supercharging. The benefits of smaller capacity engines and associated problems and solutions are reviewed in the following sections.

3 BENEFITS OF SMALLER CAPACITY ENGINES

The principal benefits resulting from the use of a smaller engine are:

i Reduced engine weight with consequent savings in the weight of transmission components and body structure.

ii Reduced friction.

iii Reduced pumping losses because of operation at wider throttle openings.

iv Improved fuel economy if reduced levels of performance are accepted.

These factors will combine to improve engine efficiency at part load and WOT and also reduce vehicle fuel consumption and CO_2 emissions.

4 DISADVANTAGES OF SMALLER CAPACITY ENGINES AND SUGGESTED REMEDIES

4.1 Reduced Power

This factor, inherent in the use of smaller engines, can be addressed at two levels:

i Change of consumer expectations by education.

ii Creation of high specific power output engines.

Any modification of consumer expectations would require a coherent international total energy and emissions policy allied to different marketing and advertising strategies on the part of manufacturers.

The generation of high specific power from small engines can be achieved by intelligent supercharging where a supercharger with a bypass valve is mechanically driven through a clutch and the bypass and clutch are both controlled by the engine management system. This should have the effect of partly restoring peak performance when demanded yet giving the benefit of a smaller engine at part load operation. Results published by Toyota (1) indicate that a supercharged 2 litre engine gives similar levels of performance to a naturally aspirated 3 litre but with a improvement in fuel economy of 7 per cent.

4.2 Higher Loading

The need to operate closer to WOT and the high in-cylinder pressures resulting from supercharged operation will impose a higher loading on the engine. This would lead to accelerated wear patterns with consequent difficulties in maintaining emission levels and possibly shortened engine life. These problems could be offset by heavier construction but this would carry a weight penalty.

4.3 Effect on Emissions

NO_X levels from a small supercharged engine could be higher when maximum performance is demanded. In general the engine would not be expected to run in supercharged mode for long continuous periods so this problem may be alleviated by operating patterns.

5 CONCLUSIONS

Downsizing of engines for passenger cars would yield near-term reduction in fuel consumption but would require a change in consumer expectation of performance levels.

6 REFERENCES

1 KATO, S., NAKAMURA, N., KATO, K. OHNAKA, H. High efficiency supercharger increases ingine output, reduces fuel consumption through computer control. SAE Paper No 861392.

Automotive Heat Engine Technology Program

Richard T. Alpaugh

U.S. Department of Energy
Office of Transportation Systems

Background

In 1988, the transportation sector accounted for 27% of all U.S. energy consumption and perhaps more importantly over 63% of all U.S. petroleum usage. Since 1978, the amount of crude oil used in transportation has exceeded all domestic crude oil production. By 1988, petroleum consumption by the transportation sector was 23% more than crude oil produced in the U.S.

The petroleum shortages of the 1970's prompted legislative action by Congress addressing various conservation activities. Recognizing the significance of the transportation sector as a petroleum consumer and the technical difficulties of achieving energy conservation in mobile equipment, Congress passed the Automotive Propulsion Research and Development Act in 1978 to address the transportation energy issue. The act, which states that "It is the purpose of Congress to direct the Department of Energy to make contracts and grants for research and development leading to the development of advanced automotive propulsion systems...", is the genesis of the present Heat Engine Propulsion Program which, broadly stated, is intended to enhance energy security, reduce oil imports with their related trade payments and encourage economic growth by improving transportation system energy efficiency.

Within the transportation sector energy use is dominated by the automobile which accounts for 45% of all petroleum consumption. (Fig. 1) Other highway vehicles including heavy and light trucks bring the highway consumption to over 80% of the total transportation usage. The remaining 20% includes marine, rail, and air. Because of the high energy use in autos and trucks, highway systems have become the principal target of conservation research and development activities in DOE.

From an economic perspective in 1985, expenditures for gasoline for all U.S. automobiles exceeded the expenditures for the purchase of new cars. A one MPG improvement in auto economy could effect a cost savings of $4.3 billion per year, while for comparison, a 1% reduction

OIL USE BY MODE, 1988
(PERCENT)

21.3 QUADS

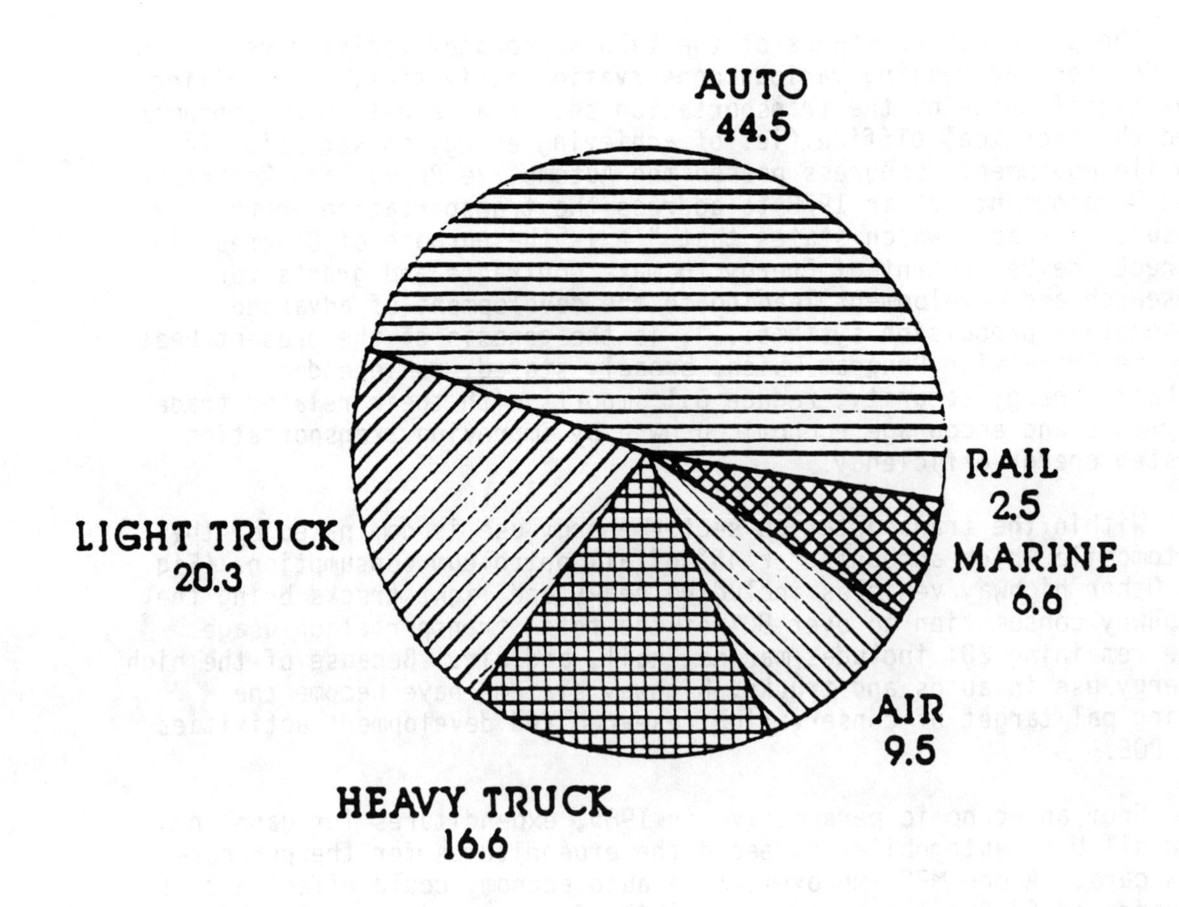

FIG 1

in new car costs could save $870 million per year (Fig. 2). Clearly the benefits of conservation in transportation are significant from both an energy and an economic viewpoint. In addition to energy conservation, air quality continues to be a major national concern likely to grow in importance with time. CO^2 emissions from transportation vehicles contribute significantly to Greenhouse problems. If Vehicle Mile Traveled (VMT) growth continues at its historical rate of 3% annually, large increases in fuel economy or other effective measures will be required just to maintain present CO^2 emission levels.

Program Goals & Objectives

The goals of the Heat Engine Programs are to improve end-use efficiency and to expand the capability for substitution of alternative fuels for petroleum by developing an applicable industrial technology base. To achieve these goals a balanced program of high risk, high pay-off R&D on promising transportation heat engine propulsion conservation technologies has been constructed. This program aims to provide the automotive industry with proof-of-concepts for advanced engine technology with potential for fuel economy improvement on the order of 30% while meeting or bettering emission standards. Similarly the advanced diesel technology efforts are expected to provide the heavy duty transport industry with like fuel economy improvements. Supporting technologies are included in the program with emphasis on the development of a fundamental advanced materials technology base to provide the ceramics industry with the capability of producing reliable and cost effective components for advanced heat engines. A utilization-oriented fuels data base for heat engine operation on biomass, oil shale and coal derived fuels is also in development in the program.

Major Program Areas

The DOE Heat Engine Propulsion Program is divided into two major program areas- Engine Systems Development and Engine Technology Development. Three sub-programs - the Automotive Gas Turbine, the Automotive Stirling and the Advanced Heavy Duty Diesel - are included in the Engine Systems Development Program. The Ceramic Technology for Advanced Heat Engine Program and the Alternative Fuels Utilization Program are elements of the Engine Technology Development Program.

Automotive Gas Turbine

The Automotive Gas Turbine (AGT) program was conceived with the goals of achieving a 30 percent improvement in fuel economy over contemporary spark ignition engined automobiles of equal performance and curb weight, improving the environment through lower CO and HC emissions and providing enhanced alternative fuel capability. To achieve these goals a proof-of-concept approach with experimental engines was taken with two industry teams representing both the gas turbine and automotive industries. The primary technical challenges

1985 EXPENDITURES FOR NEW AUTOS AND GASOLINE

POTENTIAL SAVINGS FROM TWO OTS ACTIONS COMPARED WITH OTS BUDGET

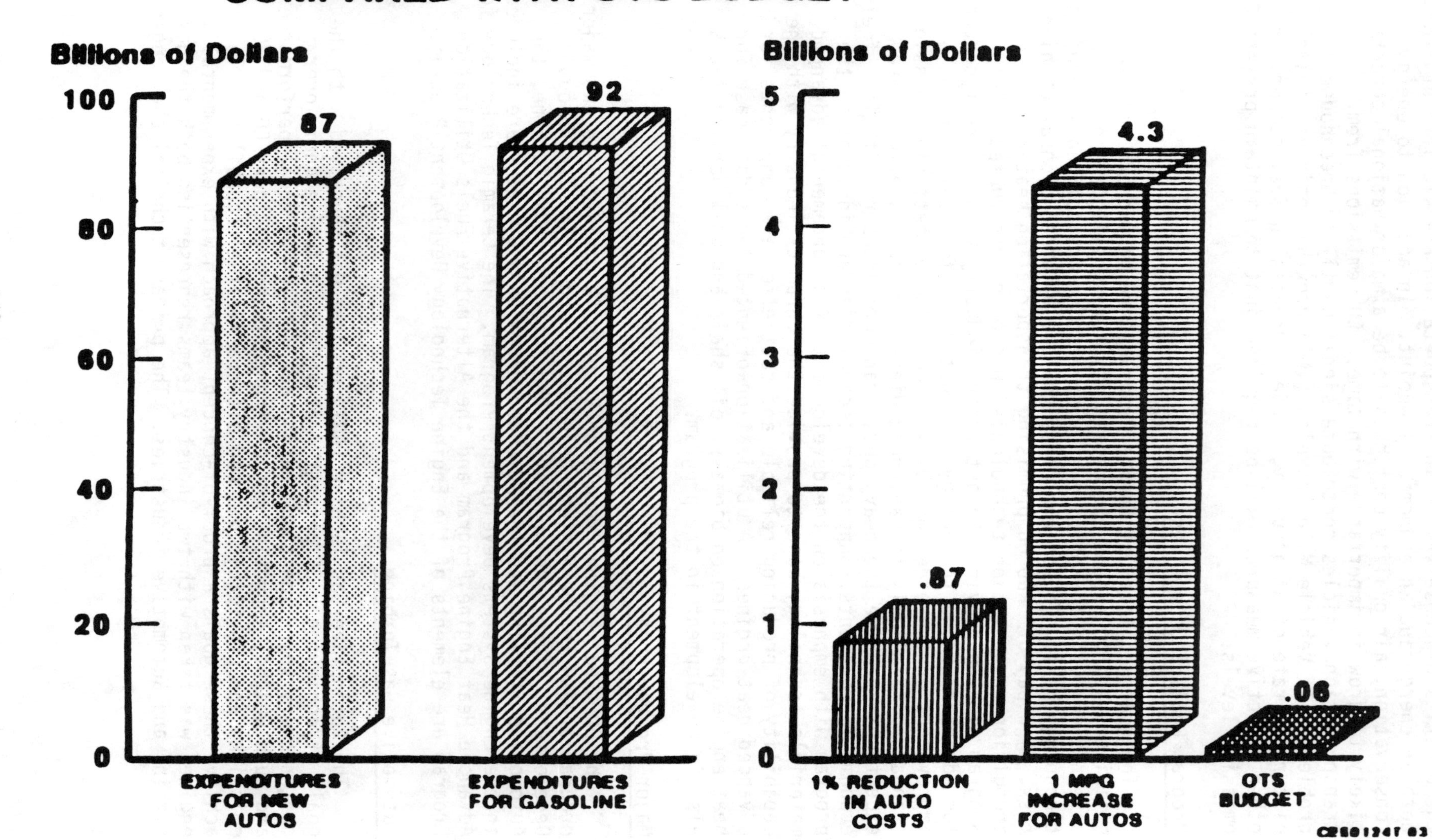

faced included the development of small, efficient turbomachinery, operation at turbine inlet temperatures in the 2350°F to 2500°F range, the development of reliable high temperature structural ceramic components and the development of low emissions combustor technology.

The following are among the more significant accomplishments thus far achieved in the AGT program.

- o Two different engines were designed (the AGT100 and AGT101) and power trains were fabricated and successfully tested at near design conditions of RPM, turbine inlet temperature (TIT) and horsepower.
- o Engine operation with all-ceramic hot-flow-path components was performed for 85 hours at a TIT of 2200°F.
- o Proof testing of 137 components in a test bed engine was accomplished and structural ceramics passed over 40 hours of exposure to 2500°F.
- o Emissions goals were demonstrated using diesel fuels and methanol in test-bed engines and low emissions were demonstrated with JP-5 fuels in a combustor rig.

Despite the progress made in the program. it became apparent that high temperature structural ceramic development had not progressed to the point that fully reliable ceramic components could be designed and fabricated for turbine application. As a consequence the gas turbine program has been re-oriented to provide a greater focus on ceramic application. This new program called the Advanced Turbine Technology Applications Program (ATTAP) has as its objectives the establishment of reliability of ceramic component designs and materials, the expansion of the experimental materials data base in an operating environment and the development of the analytical tools needed to support industry in the successful application of ceramics to long-life turbine engine. ATTAP will be closely coordinated with DOE's Ceramic Technology for Advanced Heat Engine Program and new and improved materials and technology emanating from that program will be applied where appropriate.

Automotive Stirling Engine

Program goals for the Automotive Stirling Engine (ASE) Program are similar to those for the Gas Turbine with fuel economy improvements on the order of 30 percent, expanded alternative fuel capability and improved emissions being the primary targets of the program. The project began with early proof-of-concept testing of a stationary Stirling engine adapted for automotive installation. Successive engine models were developed to improve fuel economy, vehicle acceleration and engine weight/volume. The first generation MOD I and the upgraded MOD IA engines advanced Stirling Technology to a verified 15 percent improvement in fuel economy with acceptable acceleration. The second generation MOD II engine currently in development will be installed in a U.S. Postal Service Long Life Vehicle as a final Stirling engine technology demonstration of the project goals.

In addition to the development activity on the MOD II engine, a technology transfer program is also being supported. This program known as the Government and Industry Participation Program (GIPP) provides surplus MOD I engines for additional testing by industry and other government agencies. The government-owned engines are loaned at no charge to voluntary participants. Present participants include NASA, U.S. Air Force, Deere and Co., The American Trucking Associations and Purolator Courier.

Recent project accomplishments include:

- Full MOD II engine system characterization with over 800 hours of engine testing.
- Demonstration of multi-fuel capability of MOD I engine using gasoline, diesel and JP-4 fuels during a 1000 hour program of mission operation on an Air Force van.
- Analyses indicating favorable fuel economy, emissions and manufacturing costs for the MOD II second generation engine design.
- Installation and test of a first generation MOD I engine in an Air Force pickup truck.

Advanced Heavy Duty Diesel

The objective of this program is to develop a technology base for advanced heavy duty diesel engines used in trucks, buses and other non-highway transportation systems. Goals have been established for attaining a rated specific fuel consumption of 0.25 lbs/BHP-HR and providing an economically and socially acceptable technology meeting noise and emission standards and broad fuel capabilities while being competitive in capital and maintenance costs. The programmatic approach has been to maintain a strong industry involvement in the planning process while establishing and maintaining an appropriate mix of near term and far term technology projects.

The technical program consists of four interrelated technical disciplines. These are: thermal insulation, friction and wear (tribology), combustion, and advanced thermomechanical systems.

In the area of thermal insulation, the goal is to reduce in-cylinder heat loss by up to 80% To achieve this reduction the use of thick thermal barrier coatings, solid ceramic inserts, air gaps and hybrid systems combining these approaches are being considered. Programs for the development, design application and test of thick thermal barrier coatings and solid ceramic inserts are presently being supported.

Activity on friction and wear has concentrated on the development and verification of high temperature lubricants capable of operating with top ring reversal temperatures up to 500°C. Both liquid and gas phase approaches are being pursued. Additionally, ion-implementtation low friction and wear protective coatings have been tested.

The primary concern in the combustion area has centered on high temperature emissions. Increasingly restrictive Federal emission standards make it imperative that any new diesel technology emanating from the DOE program contribute toward meeting the standards. Several efforts to characterize emissions for the low heat rejection concept are currently underway.

Approaches to advancing diesel thermomechanical systems include as a primary element the development and application of waste heat utilization concepts. The need in this area is to develop an approach that is both effective in reducing fuel consumption and economic in its back pay period. Studies of organic and steam Rankine cycle bottoming systems, diesel/Rankine cycle integration and turbocharger/ turbocompound systems are presently being conducted.

Ceramic Technology for Advanced Heat Engines

The Ceramic Technology for Advanced Heat Engines (CTAHE) program, conducted through Oak Ridge National Laboratory, was established to develop an industrial technology base capable of providing reliable and cost effective high-temperature ceramic components for application to advanced heat engines. It consists of balanced research effort in three areas: (1) materials and processing, (2) data base and life prediction and (3) design methodology.

The objective of the materials and processing research is to develop processing methods that can yield materials with uniform microstructures and high temperature mechanical corrosion-resistant components. Principal research areas include powder synthesis and characterization (to increase uniformity and controllability of properties); processing, characterization and densification of green-state ceramics; and structural, mechanical, and physical properties of dense ceramics.

Research efforts in the data base and life prediction area are focused on the development of techniques, generation of concepts, and acquisition of data on both existing and new ceramic materials necessary to improve mechanical and environmental reliability. They encompass structural qualification procedures and testing for ceramic engine components, making full use of available industrial test rights for both turbine and diesel engines; detailed study of time-dependent and environmental effects in simulated service environments (in

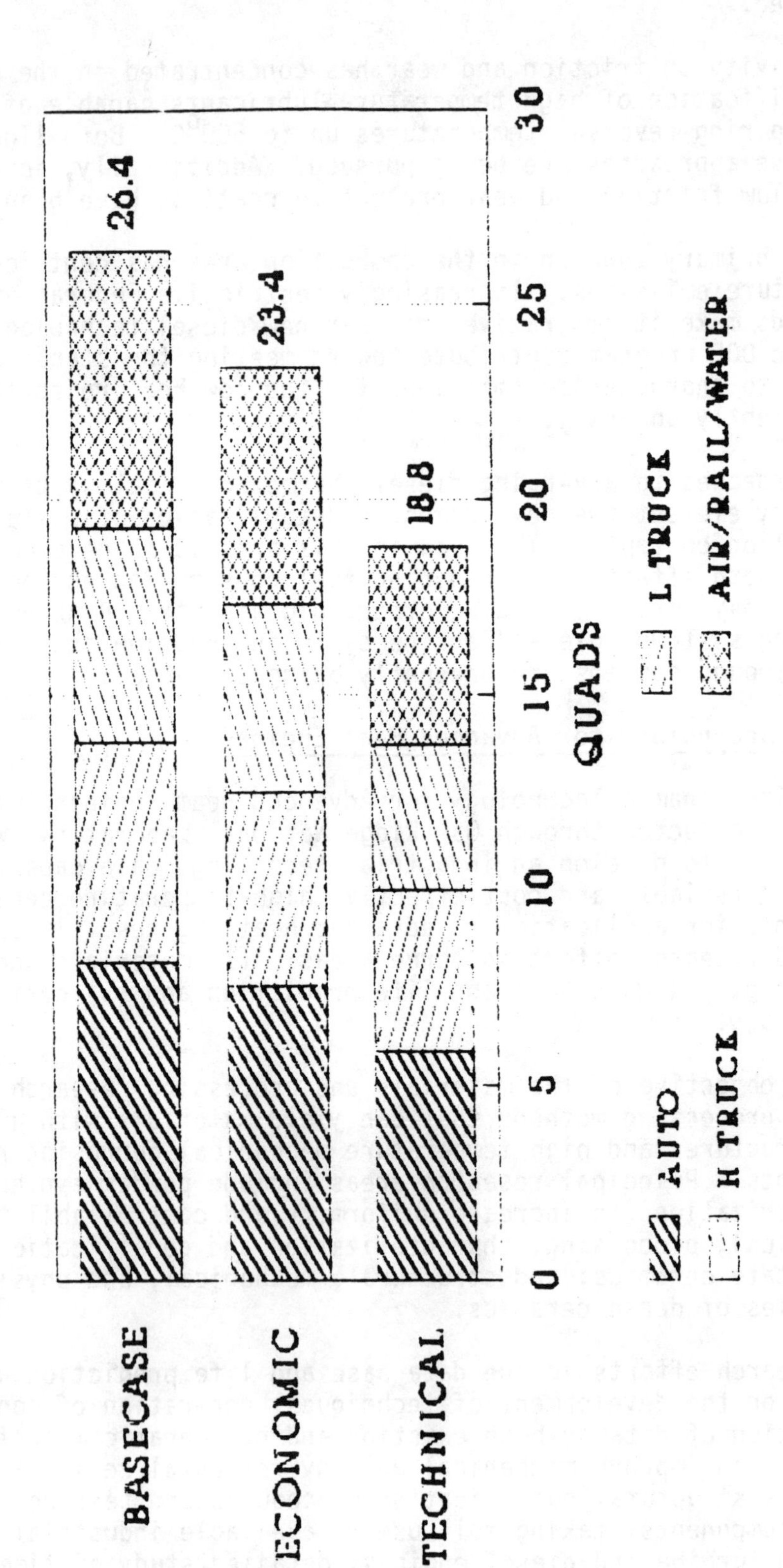

TRANSPORTATION OIL USE
BY MODE IN 2010 (QUADS)
BY CONSERVATION SCENARIO
BASECASE
ECONOMIC
TECHNICAL
26.4
23.4
18.8
0
5
10
15
20
25
30
QUADS
AUTO
H TRUCK
L TRUCK
AIR/RAIL/WATER
MYP

particular, the characterization of the long-term behavior of both existing and new materials); and development of advanced non-destructive evaluation (NDE) methods that can be used in combination with design methodology to generate accurate service-life predictions.

Research in design methodology is oriented toward understanding the properties of brittle ceramics under stress well enough to predict accurately their lifetimes in specific engine-component uses. It involves finite element modeling on a microscopic scale, studying static and moving interfaces between ceramics and other ceramics or selected metal alloys, and developing and utilizing advanced statistical representations for new design methodologies tailored to the use of ceramics in advanced heat engines.

An IEA Cooperative Program on Ceramics for Advanced Engines is currently in effect with Sweden, the Federal Republic of Germany and the United States the participants. A new agreement including Japan in addition to the past participants is in negotiation.

Alternative Fuels Utilization

The use of alternative fuels in place of conventional gasolines and diesel fuels is a means of significantly reducing the nation's dependence on petroleum. To a limited extent, alcohols (mostly ethanol from agricultural sources) blended in gasoline, and natural gas (in a very small number of vehicles) are already displacing petroleum-derived fuels. For the longer-term, it would be desirable to derive transportation fuels from abundant domestic resources such as coal and oil shale.

The primary goal of the Alternative Fuels Utilization Program (AFUP) is to assure the availability of technology for, and eliminate barriers to, the use of alternative transportation fuel options so that industry can bridge temporary and long-term gaps between petroleum supply and demand, and reduce the nation's dependence on petroleum imports by using abundant indigenous resources.

Six classes of fuels are currently addressed in AFUP: (1) new hydrocarbon fuels; (2) synthetic gasoline and distillate fuel (fuels in this category meet current fuel specifications); (3) alcohol fuels; (4) advanced fuels; (5) emergency fuels; and (6) methane and related gaseous fuels. Limited work on advanced fuels (hydrogen) has been carried out part way, consistent with the long-term prospects for supply.

The Alternative Motors Fuels Act was enacted by Congress in 1988 to encourage the development production and use of alcohol and natural gas transportation fuels. Program elements now being planned will involve a Federal light-duty vehicle demonstration, a heavy duty truck commercial application project and a bus project. Implementation will begin in 1990.

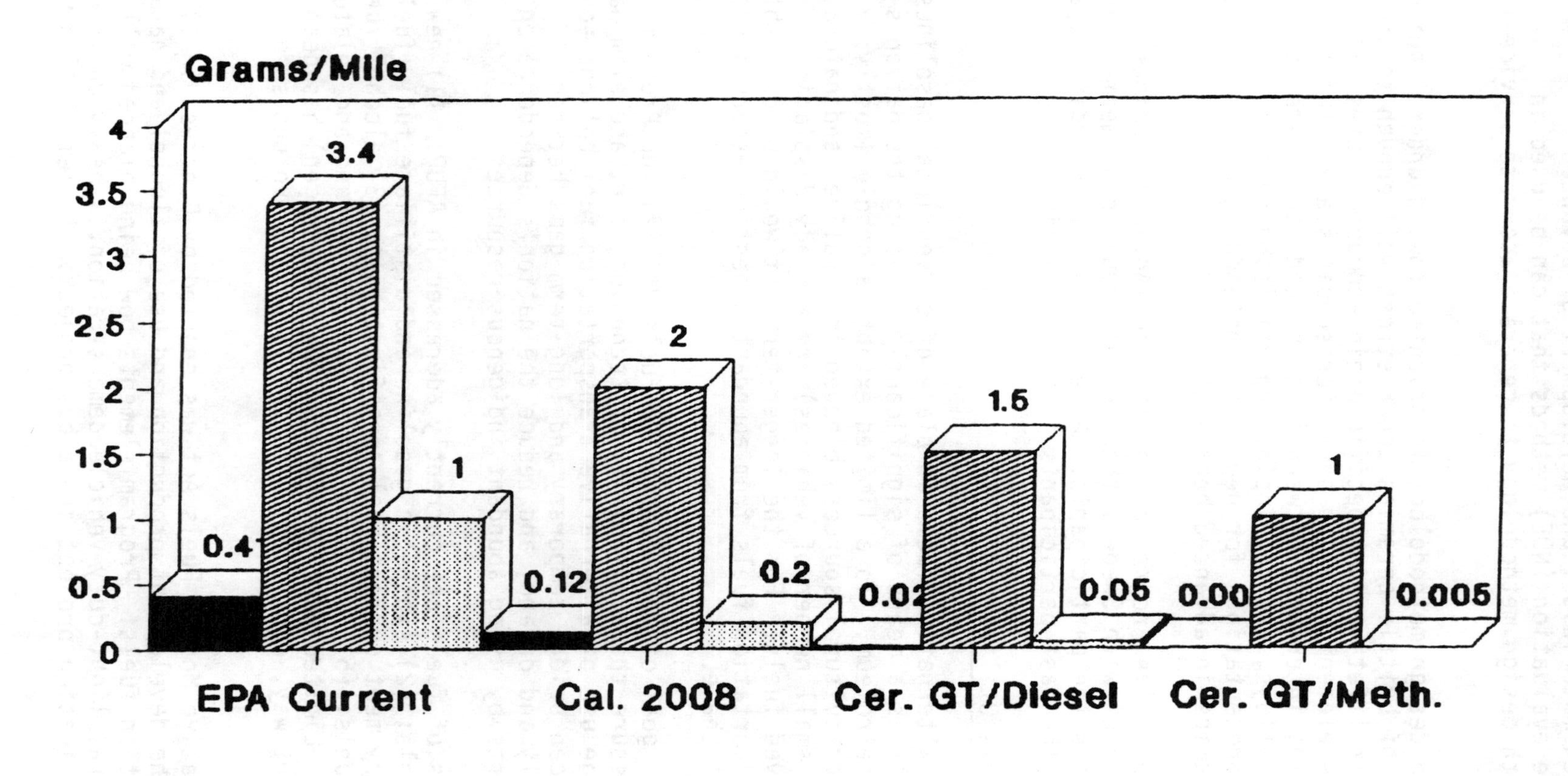
REGULATED EMISSIONS
PASSENGER CARS
Grams/Mile
4
3.5
3
2.5
2
1.5
1
0.5
0
0.4
3.4
1
0.12
2
0.2
0.02
1.5
0.05
0.00
1
0.005
EPA Current
Cal. 2008
Cer. GT/Diesel
Cer. GT/Meth.
HC
CO
NOX

FIG 4

Future Projections

Energy Use in Transportation

Many projections of future energy use in transportation systems in the United States have been made. Typical of these is a recent analysis by the Argonne National Laboratory which shows total energy consumption in transportation growing from about 20 quads per year in 1985 to over 26 quads in 2010 despite projections of growth rates in auto vehicle miles traveled only half those experienced in the past 20 years. Improvements in transportation system efficiencies can provide a significant impact in restraining this growth. Fig. 3 represents the oil use base case in 2010 by transportation mode and also illustrates potential oil savings through the adoption of advanced technologies and conservation practices as projected in a recent delphi study. The bar labeled "Economic" shows energy use if economically viable conservation options are employed. The "Technical" bar represents technically achieveable options. By this analysis, progress in conservation technology could arrest the growth in transportation oil consumption and bring 2010 levels of oil use back to 1985 levels while permitting reasonable growth in passenger and freight vehicle miles traveled.

Emissions in Transportation

Versions of the Clear Air Bill in Congress show proposed levels of automotive controlled emissions being reduced by the mid - 1990's. Unburned hydrocarbon levels would be cut from the present level of 0.41 gpm to 0.25 and NO_x would be reduced from 1.0 gpm to 0.4. Proposed emission standards for heavy duty truck and bus engines also show reductions in NO_x and particulates by 1994. Engine emission technology improvements will of course be required to meet these new standards. Fig. 4 illustrates the emissions for one such emerging engine technology, the ceramic gas turbine. As can be seen, the projected emissions for the gas turbine with both diesel and methanol fuels could meet the proposed EPA and California standards.

The Greenhouse effects of carbon dioxide engine emissions while unregulated are also of increasing concern. Fig. 5 shows normalized carbon tailpipe emissions for current technology S.I. engines, alternative fueled engines and advanced power systems. Since CO_2 emissions are closely correlated with thermal efficiencies, the advanced engine concepts designed to achieve energy conservation also provide improved CO_2 emissions.

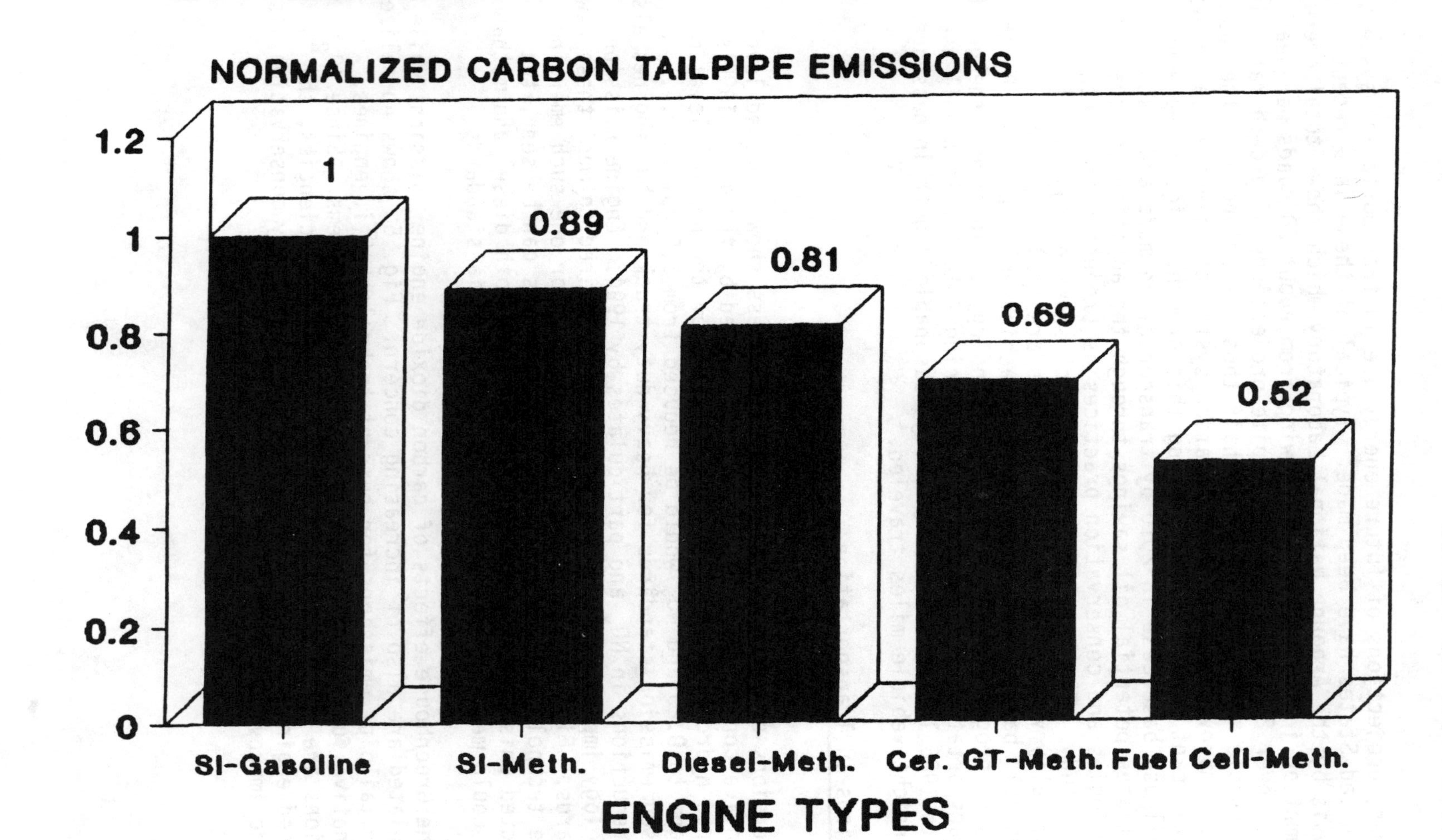
CARBON DIOXIDE EMISSIONS
VEHICULAR FLEET WITH METHANOL FUEL
NORMALIZED CARBON TAILPIPE EMISSIONS
1.2
1
0.8
0.6
0.4
0.2
0
1
0.89
0.81
0.69
0.52
SI-Gasoline
SI-Meth.
Diesel-Meth.
Cer. GT-Meth.
Fuel Cell-Meth.
ENGINE TYPES
FIG 5

Areas of IEA Cooperation

As illustrated in the foregoing sections, projected technology advances in automotive heat engines could significantly impact future levels of energy consumption and emissions. International cooperation on technology is at present being supported under two IEA agreements: (1) Research and Development on High Temperature Materials for Automotive Engines and (2) Alcohol and Alcohol Blends as Motor Fuels.

Both of these programs represent key technology areas in which future, perhaps expanded, cooperation is clearly warranted. Other subjects such as automotive and heavy duty truck emissions standards, test procedures, and engine data may also be fertile areas for future IEA cooperation.

PART IV:

OVERVIEW OF NATIONAL AND INTERNATIONAL PROGRAMMES AND ANALYSES

COMBUSTION R AND D PROMOTED BY THE CEC

Angelo Rossi

Commission of the European Communities

Outline of presentation

1) Motivations of CEC intervention in R and D for low-consumption low-emission road transportation (mainly combustion and engines)
2) Chronology of actions (4 runs)
3) General features of the 3rd and 4th runs
4) Outline of the 3rd run (1985-88)
5) Outline of the 4th run (1989-92)
6) Concluding remarks

MOTIVATIONS OF EEC INTERVENTION IN RD FOR LOW-CONSUMPTION LOW-EMISSION ROAD VEHICLES

(mainly combustion and engines)

Importance and priority of combustion RD:

-about 90% of the energy used in EEC countries is generated by combustion

-hydrocarbons, and in general fossil fuels, constitute a precious source of limited size, and have to be consumed in the most efficient ways for specific utilizations

-combustion is the main cause of atmospheric pollution

In particular, combustion in engines for road trasport:

-transport utilizes in EEC countries 40% of total liquid fuels

-very little hopes in the near future for significant alternatives

-multiplicative effect, due to the large scale of vehicle production and utilization (provided the improvements are promptly applied)

CHRONOLOGY OF ACTIONS (4 RUNS)

-Early initiatives

Runs 1976-1979 and 1979-1983: scarce funds, only scattered activity, no critical size for effective coordination

-Run 1985-88

Dangerous two-years interruption

More money available for systematic approach

Harwell study 1985, coordination meetings

Approved programme included all major priority items in combustion RD, from basic-oriented research in advanced diagnostics, chemical kinetics to 3D models for IC engines

EEC contribution about 8.5 MECU, plus equivalent industrial contribution

About 50 partners in 12 contracts

-Present run 1989-1992 (JOULE)

Approved beginning 1989 for 39 months

Call for proposals closed april 1989

Only 7 MECU available for combustion RD

56 proposals received, requesting 43 MECU

Concentration on IC engines

More integrated approach: energy saving and environment

OUTLINE OF 3rd RUN (1985-88)

1) Basic combustion RD
   - -Turbulent combustion and diagnostics
   - -Kinetics and elementary chemical processes
2) Development and validation of simulation models for IC engines (Otto and Diesel)
   - -model for fuel spray
   - -model for homogeneous combustion and knocking
   - -general model for reciprocating engines, incorporating the previous models
3) Improvements in IC engines
   - -Otto cycle engines: final phase of the cycle
   - -Diesel engines: two projects, aiming at improving performance and understanding behaviours (efficiency and pollution)

GENERAL FEATURES OF THE 3rd AND 4th RUNS

-focused on oriented-basic research and development, with industry setting the long-term objectives

-based on close interaction between industry and universities and research institutes

-close interactions among "complementary" contracts, with contractual procedures for exchange of essential information

-basic projects have university or research institute as project leader

-more applied projects have industry as project leader

-dissemination of non confidential information in reports, symposia, seminars, and publication in international magazines.

-active interest and participation by non-EEC countries

OUTLINE OF THE 4th (JOULE) RUN (1989-92)

(not yet complete)

Basic research on diagnostics

- Multidimensional imaging, Laser Induced Fluorescence

Diesel engines

- IDEA (Integrated Diesel European Action): complete modelling and validation of Diesel combustion
- Mixture formation parametres important in Diesel combustion (squish, wall efeects; models and validation)

Otto engines

- Study of the final phase of the cycle, to improve operation, especially in part load regimes
- Damaging mechanisms in knocking combustion

TENTATIVE CONCLUDING REMARKS

Responsibility of panels of experts like this one for effective and good recommendations

Technical soundness of proposed regulations (need of pre-standardization research)

Economic feasibility of proposed regulations

Danger of gaps in financing RD

Danger of relaxation

Need of political, financial, regulatory,.... actions to quickly (and automatically) implement technical improvements on a large scale

Opportunity of a fuller integration (after energy plus environment, expand to a wider time-scale over the life-cycle of the product)

CEC RD PROGRAMME

NON-NUCLEAR ENERGIES AND RATIONAL USE OF ENERGY

(1989-1992)

J O U L E

Joint Opportunities for Unconventional or Long-term Energy Supply

Section 2.1.2 of Request for Proposals

Combustion technology

The objective of future work is to carry out basic combustion RD, to develop diagnostic techniques, and to construct simulation models, which will serve as design tools for the development of more efficient, less polluting internal combustion engines.

This work will be carried out in collaborative projects by industry, universities and national research laboratories.

a) Development of tools and basic research needed for the development of different combustion systems:

- basic RD on kinetics and turbulence in combustion processes, including numerical modelling, and research on pollutant formation and control (soot particles, polyaromatic hydrocarbons). This research should be related to technologies such as petrol or diesel engines, combustion in industry, gas turbines, etc.
- collection of data in a data base for use in computer simulation models
- development of diagnostic techniques for measurements on combustion processes
- development of simulation models, their experimental verification and validation. These models will be used for the design of petrol and diesel engines

b) Development of energy-efficient, low-polluting and cost-effective combustion systems, such as reciprocating internal combustion engines (petro and diesel engines) for road vehicles and large diesel engines for ships and stationary applications. In particular, the problem of pollutant formation in diesel engines will be addressed.

MACRO-ECONOMIC VERSUS MICRO-ECONOMIC PHILOSOPHIES

	Energy saving Environmental protection	Money saving Profit making
Macro-economic point of view (the nations, the Community,..)	Primary objective	Secondary objective
Micro-economic point of view (the individuals, the families, the companies,...)	Secondary objective (or not at all)	Primary objective

Connection between the two philosophies

Large-scale application of technical findings

Automatic transfer to money saving through careful, stable, uniform, harmonized set of rules by public authorities.

Pure uncorrected market forces not always sufficient

ENERGY SAVING AND ENVIRONMENTAL PROTECTION

The need for a fully integrated approach

- a certain integration already present in the 3rd and 4th runs
- more integration needed on a wider time scale: considerations on the life-time total effects (energy consumption, environmental effects, and economic aspects over the whole life time of the product, including design, manufacturing, materials supply, marketing, distribution, operation, maintenance, final disposal and/or recycling)
- long-term analogy between cash-flow and energy flow: opportunity of values actualization in both cases for big long-duration projects (typically power stations, but why not the abandoned concept of long-life, say 20 years, car?)

Traffic and environmental policy in the Netherlands

Martin Kroon (Department of the Environment, The Netherlands)

Introduction

Never before in a Western European country has a government fallen over an environmental issue, let alone a question related to the reduction of car use. Yet on May 2nd 1989 the seven year-old Lubbers Government, being a coalition of centre Christian-Democrats and right-wing Liberals, split over the question of curtailing tax benefits for (car)commuters in the Netherlands. Surprisingly, it was not the Cabinet itself that split. After weeks of interministerial negotiations the Cabinet itself had finally reached agreement on a Dfl. 650 million commuter tax reform that would create a funding basis for several environmental and public transport programmes.

But the mere rumour of a tax reform caused great concern among the Liberal party's members and representatives. So it was that just one of the great many projects under the new National Environmental Policy Plan 1990-1994 caused a political crisis for a largely successful coalition.

It is worthwhile examining what environmental problems and traffic policy concerns are at stake in the Netherlands, and which measures are being developed under environmental and traffic policy responsibilities. This contribution describes current policy developments in the Netherlands for the reduction of pollution due to motorized traffic.

Traffic and environment

In the Netherlands, transport and communication activities represent more than seven per cent of gross national product, even surpassing the agriculture sector in economic output.

The negative effects of road transport activities mainly include accidents, congestion, air pollution and noise, wastes, soil pollution by spilled fuels, energy consumption and consumption of land, space and other resources for infrastructure and vehicle use. The non-internalized social costs of road transport probably amount to several per cent of the gross national product. Emissions from the transport sector represent a high share of overall man-made emissions. Also, the contribution of the transport sector in the total emission of air pollutants and noise is higher than in the past, compared to the contribution of other sectors.

Road traffic is the largest single source of air pollution and noise nuisance. More than six million motor vehicles travel a total of about 100 billion kilometres a year (1988), producing 723,000 tons of carbon monoxide (CO), 198,000 tons of hydrocarbons (HC), and 299,000 tons of nitrogen oxides (NO_x). Its CO_2-output (25 million tons) represents 15% of the Netherlands' contribution to global warming. Furthermore road traffic is by far the largest source of environmental pollution in urban areas, not only for the compounds mentioned above but also for particulates, asbestos, SO_2, and noise nuisance.

Table 1 shows the volume of air pollutants that are predominantly produced by road traffic in the Netherlands.

Table 1: Emissions road traffic (NL), in tons per year and percentage

	(1970)	1988	
1. Lead	950	340	80%
2. CO	1.360,000	723,000	70%
3. NO_x	134,000	299,000	55%
4. HC	200,000	198,000	45%
5. Asbestos		480	35%
6. Particles		36,000	20%
7. CO_2		25.000,000	15%

The effects of vehicle emissions can be distinguished in those concerning human health and those affecting the environment as a whole. Those effecting human health are:

a. **nuisance**: noise, odour, haze and decrease in visibility due to mild smogs;
b. **irritation**: of respiratory systems, eyes, skin, etc. by nitrogen oxides, sulphur oxides, oxidants, particulates;
c. **toxic systematic action**: carbon monoxide, lead compounds, certain hydrocarbons;
d. **mutagenic/carcinogenic action**: particulates, asbestos and certain hydrocarbons (polycyclic aromatic hydrocarbons, dioxins, benzene).

Recently, hot and still weather in Mexico City, Athens and Barcelona has demonstrated the lethal effects of road traffic air pollution. High concentrations of these air pollutants are mainly found in urban areas and near busy motorways. High ozone concentrations due to transboundary pollution and domestic traffic emissions occurred several times in the Netherlands in the hot spring and summer of 1989.

Apart from the widespread ecological damage and general land use effect, the long term/long range environmental effects of road traffic may well be illustrated through its share in acidification and photochemical air pollution (ozone formation). In the Dutch situation, road traffic contributes substantially to both forms through its share of over 55 per cent in NO_x- and of 45 per cent in HC emissions (table 1). Acidification and photochemical oxidation are transboundary phenomena with strong damaging effects on nature, forests, agriculture and man made goods, such as monuments and archives. What goes up, must come down: more than 50 per cent of acid deposition on Dutch soils and waters originates from abroad, and more than half of Dutch $SO_2/NO_x/NH_3$-emissions is raining down in Scandinavia, Germany and over the North Sea.

Today there is enough scientific evidence that both pine- and deciduous forests in the Netherlands are among the most severely threatened by acidification in Europe. Despite its green and healthy image the Netherlands can be said to be the most heavily polluted country in Western Europe. The Netherlands is, in the industrialized world, number one in terms of population density and car density, energy consumption and conversion and agricultural production (per km^2). Chemical waste and soil pollution scandals, the pollution of the Rhine river and large scale air pollution in the Rotterdam area created considerable public awareness of environmental problems in the 1970s and early 1980s. More recently "acid rain" and the long term consequences of global warming, such as rising sea levels, brought the environmental issue to the top of public interest and concern. Contemporaneously, the Governments' environmental policy shifted towards a more effect-oriented approach, resulting in stricter emission reduction goals and a solid scientific foundation for stricter products-/process-emission standards. In 1987 and 1988 "Our Common Future" from the World Commission on Environment and Development ("Brundtland-Committee") and the report "Concern for tomorrow" by the National Institute for Public Health and Environmental Protection set the terms for a more fundamental discussion on the environment issue from a global and long term perspective. Also the Government itself started political discussions on the problems of traffic planning and environment and on far reaching emission reductions (70 - 90 per cent) for acidifying substances. The time was ripe for the environment to become a cornerstone of public policy. On May 25, 1989 the Lubbers Government issued its National Environmental Policy Plan 1990-1994 ("NMP"), which is a first step towards the implementation of "sustainable development" between now and 2010 and the strategy for a new environmental policy in the '90s. This plan will add more than 6 billion guilders a year to the costs of environmental investments and expenditures.

New emission reduction targets and abatement policy

New and stricter reduction goals and abatement measures concerning acidification and all "contributing" sources have been laid down in the NMP. Total acid depositions (averaging 5,200 acid equivalents per hectare per annum) is to be reduced in the long run to 400 to 700 equivalents in order to prevent any ecological damage. This implies emission reductions (for SO_2, NH_3, NO_x and HC) of a magnitude of 70 to 90 per cent, which are goals that cannot be met by the year 2000. So, a set of maximum achievable emission reduction targets has been laid down in the NMP for the year 2000, resulting in 50 to 80 per cent reductions compared to 1980 emissions (table 2). Together with parallel reductions of transboundary sources (especially from Germany) this may result in an average yearly deposition of 2400 equivalents in 2000. Sadly enough, this will only slow down the continued dying of the Dutch forests and the continuation of other forms of damage. Eighty per cent of Dutch forests will still be in danger!

Table 2 emission reduction targets in NMP
Not only acidification but all kinds of pollution are dealt with in the NMP. The following emission ceilings and targets have been set for the traffic and transport sector:

Table 2

	1986	2000	2010
NO_x passenger cars	163,000	40,000(-75%)	40,000(-tons)
NO_x lorries,buse	122,000	72,000(-35%)	25,000(-75%)per HC
passenger cars	136,000	35,000(-75%)	35,000(-75%)year HC
lorries, buses	46,000	30,000(-35%)	12,000(-75%)
CO_2 total road traffic	24,000,000	24,000,000(0)	21,600,000(-10%)
Noise passenger cars	80	74	70
Noise Lorries/buses[3]	81-88	75-80	70
Noise nuisance serious[4]	260,000	130,000(-50%)	
Noise nuisance to any degree[5]	2,000,000	1,800,000(-10%)	1,000,000(-50%)

[3] target values for the maximum noise production of vehicles in dB(A).

[4] number of dwellings exposed to an unacceptably high noise level, reduced by 50 per cent in 2000 through measures at source and in the transmission zone.

[5] dwellings subject to noise loading of more than 55 dB(A).

Additional and stricter reduction target are to be expected under the new Government coalition of Social-Democrats and Christian-Democrats. The new Government is to bring out a supplementary plan by May, 1990, and committed itself to reduce total CO_2-emissions within four years by eight per cent. For road traffic this might imply a nett cut-off of the expected autonomous growth of CO_2-emissions between 1989/1990 and 1994.

Other goals and objectives

The use of carcinogenic or other harmful substances in vehicles must be reduced by the year 2000 to a level where the risks are negligible, and the quantity of reusable materials must be raised to 85 per cent.

In terms of land use, further "scatteration" in rural areas will be prevented. If new communication links are absolutely necessary, compensatory measures will be taken where possible so that, on balance, fragmentation does not increase. The problem of soil and air pollution at petrol stations will shortly result in legislation on new and existing facilities regarding sanitation, vapour-return etc.

The policy conducted will be regularly checked to see whether it is effective. Calibration points are given for 1994 to see whether the reduction in the environmental impact in the period prior to 2000 is proceeding according to plan. If the calibration point for 1994 is not achieved this will result in the timely preparation of supplementary policy.

Table 3

Calibration points in the NMP for emissions policy *

	1986	1994	2010
NO_x passenger cars	163,000	100,000	40,000
NO_x lorries, buses	122,000	110,000	72,000
CO_2 total road travel	24,000,000	26,400,000	4,000,000
Noise nuisance serious	260,000	190,000	130,000
Noise nuisance to any degree	2,000,000	1,900,000	1,800,000

*The units are the same as in table 2

Main abatement policy lines

Specifically, the objectives of the NMP for traffic and transport have been formulated as follows:

- vehicles be as clean, quiet, economical and safe as possible and made of parts and materials which are optimally suitable for reuse;
- the choice of mode for passenger transport must result in the lowest possible energy consumption and the least possible pollution. On the basis of anticipated technical developments this means a preference for public transport and bicycles for the coming decades. Great attention must also be paid to reducing energy consumption and environmental pollution in freight transport;
- the locations where people live, work, shop and spend their leisure time will be coordinated in such a way that the need to travel is minimal.

Environmental pollution by road traffic is produced in a 3-step process, involving (1) the vehicle emission factor, (2) the "automobility" volume factor, and (3) the traffic/driver factor. The Dutch environmental policy towards road traffic is set up along parallel lines (see Figure 1).

Figure 1

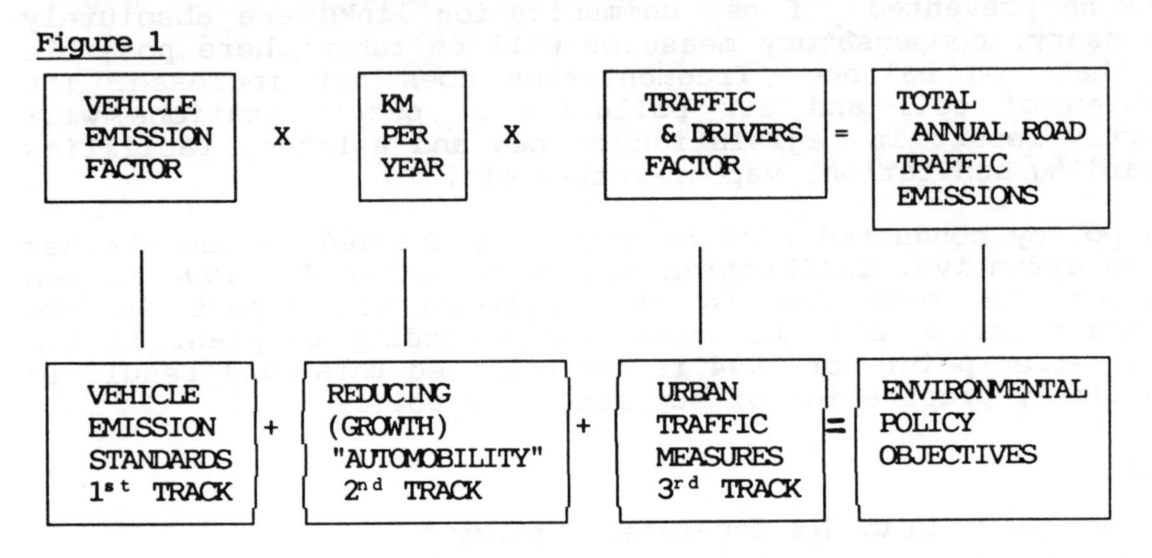

Technical vehicle standards (the first track)

The first track approach is followed throughout the world as a natural and effective means of reducing car pollution "at source" through regulations limiting air pollution and noise per (new) vehicle. Through regulations limiting air pollution step by step over a certain period of time, car makers have been persuaded to start research and produce cars with catalytic converters and electronic engine management that emit up to 90 per cent less air pollutants than similar cars in the past.

Undoubtedly the US and Japan are miles ahead of the EC in regulating exhaust gases. It should be realized that the European Community -being a supranational body with 12 member states- has an almost exclusive legislative power regarding technical standards for products to be marketed within the Community. In doing so the EC establishes a harmonized regulatory framework in order to protect the free flow of products within the EC market. As a consequence EC directives mark the limits within which member states may set standards for air pollution, noise, etc. As a member state, the Netherlands participates in negotiations regarding pollution standards and tries to reach agreements on the highest possible levels of abatement and control.

In 1985 the so-called Luxembourg agreement established more stringent EC exhaust gas standards for almost all categories of passenger cars (see Table).

Table 4

engine capacity	date of implementation	gram per test CO	$HC+NO_x$	NO_x	present share
> 2000 mm^3	1-10-88/89	25	6,5	3,5	circa 7%
> 1400 and < 2000 cm^3	1-10-91/93	30	8		circa 40%
< 1400 cm_3	1-10-90/91	45	15	6	circa 53%
Present standards "15/04"		58-110	19-28		

In June 1989, again in Luxembourg, the European Council of Ministers of the Environment agreed upon even stricter standards for cars under 2 litre engine capacity as from October 1992.
Now it can be said that from that date all new cars entering the EC-market will comply to standards equivalent to current US ('83)-standards.
From the Netherlands' Clean Car Research Programme a clear view on emission performances of current engine technologies evolves in terms of grams (in NO_x) per kilometre in conformity to standard test procedures (see figure 3).
A parallel -even more interesting- picture appears from the "conformity of production" testing of vehicles-in-use, combined with a high speed cycle test and engine tuning (adjustment to manufacturer's specifications). It may be concluded that both in practice as well as in laboratory conditions the closed loop catalyst technology (with lambda-sensor) represents the state-of-the-art in air pollution reduction (see figure 4)!

Considerable progress still has to be made in a wide variety of issues to be covered by EC-standards, before all these regulations can be said to be equivalent to state-of-the-art technology. Reducing the total air pollution of road traffic requires constant screening of those factors that influence the real exhaust gas composition of all categories under all "real life" driving conditions. This year and next year new standards will be established for the following items: light commercial vehicles; a high speed test cycle and high speed emissions; diesel-emissions of particles and gaseous compounds; evaporation losses of petrol engines cars; enforcement and lifespan reduction standards for catalyst and other reduction equipment.

Measures in the Netherlands

As from April 1, 1986, several measures entered into force in the Netherlands in order to promote the introduction of "clean" cars. Regular leaded gasoline was replaced by unleaded regular, and fiscal profits were attributed to the purchase of

"clean" cars that comply with the new EC standards. As a result, today about 80 per cent of all newly sold cars are catalyst-equipped while more than 90 per cent of all new (petrol) cars meet these special tax requirements. This approach - using economic incentives to compensate for additional costs of catalyst and other emission reduction equipment - proves to be an effective way to "clean up" the passenger car fleet long before such could be done by compulsory measures alone.
In 1988 a publicity campaign on "clean" cars, catalytic converters and unleaded gasoline was organized. The campaign sought both to remove some of the misunderstandings surrounding these topics and to promote the sale of cars fitted with catalytic converters. This campaign was set up in close cooperation with the Dutch petroleum industry and car importers.

In February 1989, it was decided to adapt the fiscal regulation in such a way as to provide maximum benefits only in respect of those vehicles that comply with US '83 standards. Until the new EC standards enter into force, compensation will be provided for the purchase of "clean" cars that comply with the new standards at a rate of Dfl 1.700 for closed loop catalyst technology and Dfl 850 for open loop or any other (non catalyst) technology that complies with milder EC standards. This regulation entered into force by March 13, 1989. Only one year later, already two out of three newly sold cars was equipped up to US '83 standards!

In order to guarantee that cars do not cause more air pollution and noise than strictly necessary during their life, general periodic car inspections will be extended to cover environmental aspects. In addition, an extensive random-sample inspection program is being undertaken over a five year period to check the anti-emission devices in use, in terms of their effectiveness, durability and their conformity to production standards.

In addition to the introduction of cleaner passenger cars, the rapid introduction of cleaner lorries is also desirable. A gentleman's agreement signed with the manufacturers and importers on 29 September 1987 represents the first step in this direction, establishing a 10-15 per cent NO_x-reduction per vehicle for 80 per cent of all newly sold lorries as from 1988.

With regard to noise reduction the same incentive approach is successfully applied through subsidies for investments in lorries that meet future (stricter) noise standards.

The following actions will be involved in the National Environmental Policy Plan period (1990-1994).

Cleaner lorries and buses.

Within the EC the Dutch Government is a strong supporter of at least 50 per cent tightening up of exhaust gas standards, thus making maximum use of current technological options. The Government will also endeavour to reach agreement in the EC on a rapid introduction of cleaner vehicles if necessary by anticipating the coming into force of new stricter EC standards. For this purpose the environmental investment subsidy programme has been raised by an amount running to Dfl 90 million per year. The amount, which will be funded through an increase in diesel excises, will be raised gradually from 1 January 1990 onwards because cleaner vehicles will not immediately be available in great numbers on the European market. The programme will be terminated once stricter European norms become effective. It is expected that this policy, given the expected autonomous growth in freight traffic from 11 billion miles to 16,5 billion miles in 2010, will lead to about 30 per cent lower emissions of nitrogen oxides and hydrocarbons in 2000.

Of equally great importance is the development of even cleaner lorries and buses. In 2010 a 75 per cent reduction in the emission of harmful substances must be achieved, although the technology for this is still unknown. The Ministry of Housing, Physical Planning and Environment and the Ministry of Transport and Public Works together with the lorry industry have embarked in 1989 on wide scale international research into promising new technologies.

Transport in cities will serve as a spearhead in demonstrating and applying clean technologies. A wide range of possibilities is available, examples being alternatives fuels (natural gas), electric vehicles, hybrid vehicles, storage of braking energy, use of particle filters etc. Many of the options can be applied simultaneously. In view of the specific circumstances the optimal solution will differ from case to case. An attempt will be made in the next five years to ensure that public transport in cities is provided with clean vehicles. An annual amount has been set aside to support this development, running to DFl 30 million in 1994.
An inquiry will be made in the enforcement sphere to see how unnecessarily high emissions of smoke and soot from lorries can be prevented. An enforcement system will be developed and implemented similar to developments in Germany.

Reducing car use (automobility, the second track)
Future levels of air pollutants emitted by traffic will be determined by the average vehicle emission factor (emission per vehicle per km) and total distance travelled by all vehicles.

VEHICLE EMISSION FACTOR	X	KM PER YEAR	=	TOTAL ANNUAL EMISSIONS

The second track-approach is relatively new and will be further developed and implemented in the coming years. It follows that the expected growth of automobile use of 3 to 5 per cent a year if policy remained unchanged would inevitably

consume a large part of the emission reductions resulting from the "clean" car programme, thus frustrating environmental objectives for emission reduction on both an (inter)national and an urban scale. Indeed, growth rates of about 6 per cent would lead to actual increases of emissions, despite the successful introduction of "clean" cars. In 1988 all traffic emissions rose (NO_x by 5 per cent) compared to 1987 due to the high growth rate of passenger car and freight traffic! Furthermore the "clean" car programme cannot solve the problems of noise nuisance, land-use and CO_2 emissions that are expected to increase with traffic growth.
Considering the severity of the effects of acidification and ozone formation, which is reflected in reduction goals of a magnitude of 70 - 90 per cent, urgent action is needed. Also, it takes too long a period (from now to about 2000) before the total car population will be replaced by maximum feasible "clean" cars. This demonstrates how imperative it is to take adequate measures soon. One must also realize that the issue at stake is an adjustment in a social-economical trend, which is no easy task and should not be delayed any longer.
Thus it is that the Second Traffic and Transport Structure Scheme (SVVII) seeks a balance between individual freedom, accessibility and environment. It has been concluded that the only way of doing sufficient justice to all these aspects is to control the use of cars.
Consequently, a set of new measures and improvements in current approaches are to be developed to tackle the "automobility" problem. They may result in a drop in the growth of automobility from 70 per cent to 35 per cent in 2010, while further reductions are needed with a view of CO_2 reduction targets.

It may be deduced from the above that the volume of traffic is an independent factor the effect of which on total emission levels increases in relation to the failure to introduce stricter emission standards, or the delay involved in introducing them. In the long term, the proportion of growth in Dutch traffic volume accounted for by freight traffic assumes an even more important role. Thus, the growth factor reinforces the necessity for the EC to do all it can to minimize the vehicle emission factor, the more because all EC member countries expect high growth rates until at least 2010.
The second track brings about a fundamentally different approach from the "classical" approach in both traffic and environmental policy. However, even for reducing traffic jams a substantial reduction of car use is thought to be justified and effective. Thus both environmental and traffic policy goals may be reached simultaneously through the same instrument, reduction of car use.

In 1987, a specific task-force, supported by McKinsey & Co., developed a series of economic and traffic measures to curb congestion in the crowded Randstad of Holland, not only by constructing new infrastructure, but also by raising the cost of car use and parking. Several other measures may pull car drivers into public transport or even on to the bicycle saddle. This "Randstad Mobility Scenario" report provoked much applause from some interested parties but even more criticismfrom car-related businesses and driver organizations. Undoubtedly, any substantial reductions of traffic volume can only be realized when both Parliament and society as a whole are willing to change the balance of interest between environment and unlimited mobility.

Limits to the reduction of car use

To what extent may car use - or at least the growth of total mileage - be limited without disturbing the economy or society as a whole? Several studies show that a considerable part of car use in the Netherlands is not "essential": Indeed, an estimated 50 per cent of all car rides may be judged as having a reasonable substitute in public transport, car pooling, telecommunications or the bicycle. Nearly half of all car movements are performed within reasonable cycling distance (five km) or even walking distance (two km).
With regard to air pollution both long distance daily travelling (NO_x) and frequent, short, cold start-and-stop trips by car (shopping, commuting, social and educational visits) with relatively high CO and HC emissions should be substituted with priority.
The Netherlands is provided with the most effective and ecological answer to the needs for short distance mobility: 12 million bicycles. Moreover public transport is relatively well developed. On the other hand, a great many factors structurally favour car use: the autonomous growth in car owning/driving age-groups, the psycho-social benefits of car ownership, the physical distribution of housing and employment locations, growing job mobility, rising job participation for married women, fiscal and other incentives, and -last but not least- the inherent advantage of cars in terms of speed, convenience, privacy and freedom. Undeniably, the Dutch and their political representatives show a strong aversion to governmental interference in private behaviour, especially when such measures are unconventional or burdensome, or both. In fact, it may be expected that government measures that raise the cost of car use in order to influence present drivers' behaviour, will meet strong opposition from various parts of society. The recent government crisis proves the strength of such opposition and of its political determination.

Various measures for reducing car use

Evidently, under current social and political circumstances there are no simple measures which can have direct and major effects on the volume of traffic. This is because the guiding principles in the current decision-making process assume no limitation of car ownership; guarantees for freedom of mobility for social, business and distribution purposes; and the superiority of a market-oriented approach over regulation. Given the importance of the automobile in modern society, substantial limitation on its use requires fundamental measures, capable of influencing people's choice of mode of transport.
Only a wide ranging package of measures which complement each

other can have any significant effect. Maybe even a cultural revolution would become necessary.

Certainly, within the last three years the issue of "automobility" has developed from taboo into a political battleground and widely recognized problem. Consequently, Government papers on traffic and physical planning and environment reveal a steadily developing set of policy goals and decisions, the latest (NMP) always more elaborate than its predecessors (4th Physical Planning Memorandum, SVVII).

"Sustainable development" in traffic and transport means that a shift will have to occur in modes of transport towards modes that are less energy-consuming and less polluting. The following changes need to occur as regards choice of mode of passenger transport:
- for short distances (up to 5 to 10 kilometres) a shift from using cars to using bicycles, with a considerable increase in the number of kilometres covered by non-motorized transport;
- a shift from using cars to using public transport with as a result, twice as many passenger kilometres being covered by public transport by 2010;
- a shift from travelling by air for distances up to 1000 kilometres to travelling by high speed train.

For that purpose:
- a great improvement will be made in facilities for public transport and for cyclists as an alternative to using the car, notably for commuter traffic;
- pricing and incentive instruments will be used to influence the choice of transport mode in passenger transport;
- information and stimuli will be provided to all those involved, notably businesses, municipal authorities and individual members of the public;
- encouraging the use of public transport has to be coupled with discouraging the use of cars, otherwise no contribution will be made to sustainable development.

None of these options can be realized overnight; some of them require amendments to legislation, while the precise advantages and disadvantages of others still need to be studied carefully.
To be effective a balanced package will have to include the following elements:
* a strong increase in variable (driving) costs, possibly in combination with a reduction in fixed costs through "variabilisation", road pricing, taxation etc;
* reduced parking facilities for commuter traffic through action on pricing, volume, regulation;
* increased attractiveness of public transport by improving capacity and infrastructure, service and comfort, speed and price;
* optimum use of physical planning via concentration, and public transport orientation of land uses;
* neutralizing existing tax allowances and other financial incentives for commuters and for the use of business cars;

* cycling, education and information on mobility behaviour.

Increasing variable car costs
Variable car costs affect the use of cars while the fixed costs principally affect car ownership. If the variable costs are raised, use diminishes and it is thus an effective instrument of policy on mobility. The most direct method of achieving this is to increase fuel prices.
The level of excise duty on petrol in the Netherlands is approximately at the proposed harmonization level of the European Commission.
This proposal allows no scope for a substantial increase in excise duty on petrol. On the other hand, excise duty on diesel and LPG is relatively low or absent. Therefore costs for users of diesel and LPG cars will increase by Dfl. 540 million from 1 January, 1990 onward by an increase in the excises on diesel (of 6.3 cents per litre) and an increase in the LPG surcharge in the motor vehicle tax.
The NMP provides for a further increase of Dfl. 190 million in diesel excises with which to finance the programme for accelerating the introduction of cleaner lorries (Dfl. 90 million) and cleaner buses (Dfl. 30 million) as well as improvements in rail and waterway facilities on behalf of freight transport (Dfl. 70 million). This total of 190 million guilders corresponds to about 4.3 cents per litre of diesel.

Road pricing will be introduced as the key instrument for controlling traffic growth. It is an instrument which charges car traffic for variable costs by time and place. Road pricing will be developed and introduced as quickly as possible, at the latest by 1996. A test project will be started in 1992 in anticipation of the large scale introduction of the system. A modern electronic (in vehicle) "chip-value cart" system is being tested already.

Commuter traffic (standard tax deduction)
In order to discourage the use of single passenger private cars for commuting in favour of other forms of transportation such as car-pooling, group transport and public transport, or to encourage people to live closer to where they work, the standard tax deduction existing for commuter traffic will - in phases - be abolished and transformed into policies directed at reducing the environmental burden and improving accessibility. The new Cabinet is to formulate very soon a position regarding the allocation of the revenues generated by this measure. Amongst others continuation of the standard tax deduction for those who use or will use public transport will be examined.
Car-pooling and private group transport should receive much more emphasis because they provide for more intensive use of existing cars without a heavier burden on public transport capacity.

Increased attractiveness of public transport
Improving and extending service by public transport could include the following;

- automobile kilometre reduction and business transportation plans;
- investments in public transport infrastructure;
- investments in bicycle facilities near railway stations;
- fare and ticket integration ;
- contribution to public transport operating costs deficit;
- encouraging cooperation between transport regions;
- research and public information.

Altogether this would involve an extra Dfl. 250-275 million per year.

Kilometre reduction plan
Companies and (government) institutions will be asked to draw up kilometre reduction plans to screen commuter traffic and commercial traffic from and to the company or agency and examine all possibilities of reducing the number of vehicle kilometres. In 1990 such plans will be drawn up on a voluntary basis with a subsidy from central government.
In the long run, the drawing up of kilometre reduction plans will be part of the internal environmental responsibility of enterprises and agencies and an annual report will be drawn up giving the results which have been achieved with the kilometre reduction plan in government agencies.

Investments in public transport infrastructure
Additional investments are necessary in the urban district networks for a further improvement in the quality of public transport in the four major urban districts (Amsterdam, Rotterdam, The Hague and Utrecht).
Extra measures are being anticipated in the other urban districts to improve the regularity and stream lining of bus services and to improve halt accommodation.
Outside the urban districts a further improvement in the infrastructure for intra-regional transport will be needed.
During the planning period (until 2010) an additional investment is required to accelerate improvements to the Dutch railways infrastructure which includes doubling the amount of track, grade-separated junctions and other measures to improve speed such as TGV and "intercity-plus" trains.

Investments in bicycle facilities
The use of bicycles will be encouraged and extra infrastructural measures for bicycle traffic will be taken for this purpose. Ways of doing this are separate bicycle routes in and around towns, notably in commuter corridors, as well as routes to and from stations and bicycle parking facilities.

Encouraging cooperation between transport regions
Regional cooperation between those involved in transport and traffic, such as municipal authorities and public transport companies, is indispensable for solving the complex problems of accessibility and environment.
Central government can provide some funding if a transport region draws up a coordinated plan to solve the traffic and environmental problems in an integrated way.

The plan must aim at reducing unnecessary car traffic, improving accessibility and reducing local environmental problems. The plan may involve parking arrangements, promoting public transport, promoting bicycle traffic, integration of taxis into the transport chain, planning measures, public information and educational activities etc.
Notably, no new institutional body is envisaged: transport regions shall be developed by cooperation between existing authorities.

Improving freight transport by rail and water
In order to raise the percentage of freight transport travelled by rail and water in the future, the competitive position of these branches must be strengthened. This means giving additional attention to the infrastructural bottlenecks in the rail and waterway networks. Additional money will be made available through the increase in diesel excises already described. The investments provided for in SVVII will be carried out with greater speed. The total investment involved is about Dfl. 40 million per year in the period up to 2000.
Dutch Railways has developed new plans (Rail 21 Cargo) on the strategy to be pursued to strengthen the position of freight transport by rail into the next century. Several billion guilders are to be invested in new and existing rail infrastructure in order to triple the annual tonnage transported by 2010.

The infrastructural bottlenecks in the waterways network will be resolved at an accelerated pace. About Dfl. 30 million per year is involved in stepping up the improvements planned in the SVV II up to the year 2000. The higher excise duty on diesel fuel may also result in some shift towards freight transport by rail and water.
Further research is being done into the factors which affect choice of transport and particularly into the motives of shippers in choosing a particular mode.
In the course of 1990 a detailed study will be completed into reducing the environmental impact of freight transport and into the reduction of energy consumption in this sector. The study will also examine the possibilities of restricting the number of goods vehicle kilometres by a more efficient use of vehicles.

Tightening up physical planning policy
Physical planning policy will concentrate on discouraging labour-intensive businesses and amenities attracting numerous visitors in locations which are less readily accessible by public transport. Physical planning and environmental policy instruments will be deployed to prevent buildings being constructed in unsuitable locations. The municipal authorities are being asked to view existing building plans in this light and possibly reconsider them.

Effects of measures

The mobility measures are part of a package; it is not possible to judge their effectiveness individually. The effectiveness of the whole set of measures is expressed in a curbing of the growth in the use of cars in relation to the forecast for unchanged policy and the proposed policy in the SVV II.

Trend in the number of car kilometres (index 1986 = 100)

	1994	2000	2010
SVV-II forecast (unchanged policy)	124	140	172
SVV-II policy	120	126	156
Set of measures NMP*	-3	- 6	- 8
Trend in mobility with NMP package	117**	120	148

* It has been assumed that measures can be taken in 1990.
** 1989: 116 already!

Given current know-how in vehicle technology and the trend in car use, the objectives for NO_x in 2000 and 2010 can just be achieved (with regard to passenger cars). As regards CO_2 emissions it is expected that these will diminish by 2010 by approximately 5 per cent. Thus the CO_2 objective of the NMP can not be achieved. The same refers to the reduction objectives for 2010 for noise nuisance and NO_x/HC-emissions by lorries. New stricter reduction targets are set out by the new, centre-left Government. So the growth of automobility has to be reduced even further then was scheduled in NMP and SVVII. From a CO_2 reduction point of view automobility has to come to a stand still at 1990 levels, while still growing at about six per cent a year!
Total investment in roads may diminish as a result of the lower growth in car traffic. These savings will first occur in the second half of the SVV plan period since there are currently backlogs which have to be caught up to reach the level of completion intended. Savings on municipal and provincial roads cannot currently be estimated. The new Government is expected to reduce more substantially road construction budgets in favour of public transport and bicycle infrastructure investments.

Urban traffic measures (third track)

Due to problems of noise nuisance, air pollution, visual pollution, the problem of traffic safety and lack of space, the quality of the urban environment has seriously deteriorated. This is particularly the case in the big cities, where motorized traffic is the main cause of pollution. Most of the air pollutants present at street level originate from motor vehicles (See table 6).

Table 6: Percentage of pollutants originating from motor traffic with respect to the total amount present

	CO	O_2	SO_2	Pb
Percentage in a busy street	90%	60%	30%	98%

Air pollution through carbon monoxide, lead and nitrogen dioxide mainly originate from passenger cars. In approximately over 1000 urban streets in the Netherlands with intensities of over 10,000 vehicles per day, the concentration of these pollutants is in excess of the ambient air quality standards for CO, Pb and NO_2, set in 1987. For benzene and particles the same will be true so although air quality standards are not set yet for these substances.

Excessive levels of air pollution and noise nuisance cannot entirely be eliminated by tougher emission standards alone. In addition to the above mentioned general measures designed to reduce the use of cars, the following measures would help to alleviate the problem at a local scale:

* stricter enforcement of parking restrictions;;
* traffic management influencing driver's choice of routes;
* route signs for freight through-traffic;
* traffic-dosaging on approach roads to city centers;
* publicity designed to influence local people's driving habits;
* introduction of low speed zones;
* circulation schemes to calm traffic and to spread it more everly over the road network.

This kind of approach combines environmental protection with road safety. The implementation of the policy outlined above will, in the first instance, be the responsibility of the municipalities, although some elements of the policy are more within the realm of central government. To this end the newly amended Road Traffic Act will make it legally possible to implement traffic measures solely or partly on environmental grounds.

Central government grants for municipal infrastructure will be directed more towards reducing the environmental effects, for example on the basis of traffic pollution maps.
Air pollution and noise levels are indicated on this map and can be directly associated with the traffic flows that cause them by ussing different colours, the shades indicating the seriousness of the situation. Problem areas become immediately apparent, can be localized and further assessed. The processing of input data as well as the actual drawing of such maps is computer-aided. A digitized traffic network and a traffic model form part of the system. Modifications and the effect on air pollution and noise levels can immediately be visualized until an optimal solution is found. About twenty "pilot study" municipalities have prepared such a map. For this purpose the Ministry of Housing, Physical

Planning and Environment has allocated four million guilders per year, annex to 15 million guilders per year for traffic measures.

Conclusion

Mobility is an essential requirement of our society, and so is the environment. To combine the conflicting demands of traffic and environment in the best possible way is a demanding task to which administrations at local, provincial and state level have to commit themselves. The execution of this task is not going to be easy.
In particular, it remains to be seen whether the automobility-reduction approach can be realized without the help of a new oil crisis or a new wave of environmental concerns resulting in a real culture shock. Compared to that, breaking down the Berlin Wall will after all appear to be peanuts! For the time being, four major uncertainties remain:

1) Will there be enough political support for the unpopular measures that effectively raise the cost of car use and reduce automobility?
2) Will such measures really provoke the mass behaviour response amongst motorist as necessary?
3) Can we stay away in the long run from more dictatorial ways of influencing people's choice of transport mode and of activity-locations?
4) Can we slow the growth rate of lorry-mileage through influencing the freight transport modal split, without embarrassing (the Dutch position in) international transport and distribution?

One thing is for sure: technology can solve a great many of the problems discussed above, but "behaviour" remains a key factor when traffic and the environment are to be sustainably developed into the next century. The (car) key is in our own hands .....

Leidschendam, november 1989

CONSIDERATIONS FOR A LOW CONSUMPTION/LOW EMISSION AUTOMOBILE PROGRAMME

ALAIN MORCHEOINE, Head of Transport Department.

FRENCH AGENCY FOR ENERGY MANAGEMENT

I - THE SITUATION

1) The situation in Europe is characterized by a strong increase in demand for passenger as well as freight transport.

According to CEMT (European Conference of Transport Ministers) statistics (figure 1) :

- Passenger transport has risen by 70 % since 1970
- freight has risen by 45 %

2) The distribution of transport modes (CEMT data : figures 2 and 3) also shows very clearly that demand is increasing faster for road transport (both passengers and freight).

As concerns road passenger transport, the part of private cars is rising much faster than public transport as shown in figure 3.

3) Traffic density is rising strongly in urban areas, for instance in France the energy consumption of urban road transportation (passenger and freight) has increased by nearly 90 % since 1975 (figure 4).

4) The road systems in Europe are increasingly congested ; capacity is not growing as fast as demand ; badly regulated or nearly saturated traffic is alleged to increase consumption by over 10 %.

5) The freight transportation pattern has undergone considerable change due to the evolution in economic patterns and industrial site location. Consequently, freight flows are more dispersed and just-in-time transportation practices result in smaller freight lots. This means greater mileages with smaller trucks and higher consumptions for an equivalent demand.

II - FRENCH CONSUMER BEHAVIOUR

AFME, the Ministry of Transport, and French car manufacturers entrusted to INRETS (Institut National de Recherche sur les Transports et leur Sécurité) a series of studies called EUREV about car use patterns in France. The main results of this work are :

- a considerable modification of driver behaviour : 25 % of journeys cover less than 1 km, 51 % less than 3 km (figure 5) ;

- the definition of a series of typical cycles representing different on-the-road traffic conditions(see figure 6) which can be used to compare with standard cycles and to address further issue ;

- measurement data on consumption and emissions, the latter requested by AQA (Agence pour la Qualité de l'Air), based on the newly defined cycles, both with cold and hot engine starting (for further details see contributions by Mr. DELSEY of INRETS).

This study is extremely constructive for defining a low consumption/low emissions car research programme and public policies aimed at consumer behaviour.

Some other INRETS studies show that impatient drivers can increase their consumption by 35 %.

Single-car ownership has risen from 61 % in 1972 to 75 % in 1988 ; multiple-car ownership from 9 % to 23 % during the same time ; this latter phenomenon, together with the effects of the economic crisis, has resulted in three consequences : second cars are old which entails a rise in the overall average age of the French car fleet, a delay in the penetration of new techniques on the market, and poor maintenance standards.

Other investigations show that the proportion of small car journeys (home to work, home to school, shopping...) is rising constantly.

All these circumstances have resulted in the fact that technological advances have not affected the real consumption of motor cars ; figure 7, which shows the consumption index (l/100 km base 100 : 1975), illustrates this phenomenon very well :

- the evolution of the average specific consumption of new cars shows technical progress ;

- the average specific consumption of the overall car fleet shows the market penetration of improved technology ;

- the real consumption of the fleet betrays car owner behaviour.

From all these observations, it may thus be concluded that :

- the situation is likely to be very similar in other densely populated developed countries (Europe and Japan). But the car use pattern is probably different, both as regards travelling distance and consumer behaviour, in countries such as Canada and the USA.

- transient running and cold engine running can be assumed to be much more important than usually accounted for by car manufacturers at the engine design stage.

- the choice of technological solutions may be inappropriate, namely for urban use. For instance, a vehicle equipped with a 3- way catalytic will start from cold to travel less than 1 km for one quarter of the journeys. In this case, the catalytic function would not be activated before the end of the trip and the car will consume 6-8 % more fuel than a vehicle without a catalyst, without any improvement on emissions especially CO and HC. Catalytic exhausts may not be operational at the time when they are most needed : bad driving and bad maintenance can only aggravate the problem : what nonsense !

- **It is essential to take into account real situations and not just theoretical data, when designing automobile engines and when elaborating emission standards.**

III - THE FRENCH EXPERIENCE IN THE FIELD OF LOW CONSUMPTION/LOW EMISSION VEHICLES

Two successive research programmes were launched in France in the early eighties on low consumption vehicles ; these programmes were strongly supported and funded by Governement organizations, particularly by AFME.

First-generation research programme

EVE and VERA were initiated in 1979 by AFME (AEE at the time), which provided the only Government funds. The objective was to reduce consumption by 25 % for medium-range cars (Peugeot 305, Renault 18,...) by applying technology that was already available for industrial production at that time. Expectations were widely surpassed, with a drop in specific consumption of 35 % (measured according to EEC standards).

In 1981, the PAU/PARIS field trial for the VERA programme resulted in a consumption of 2.52 l/100 km at 70 k.p.h over a distance of 800 km.

On the KNOXVILLE/DETROIT 512 mile trial run, an average consumption of 73.7 m.p.US gallon was achieved at an average speed of 54.9 m.p.h.

Second-generation research programmes

The encouraging results from the first programme led the Ministry of Industry to launch a second programme in 1981 under the name of the "3-liter programme". Among the Ministries and public organizations supporting this programme, AFME provided 50 % of the public money.

This time efforts were to be focussed on the economic range (Renault 5, Peugeot 104).

All available technology was to be applied (engine design, body design, aerodynamics, transmission, new materials) and synthetized at the vehicle level without any consideration as to cost.

The EEC standard specific consumption test on the VESTA (Renault) spark ignition version achieved 2.8 l/100 km. Field trials carried out in 1987 showed a consumption of 1.92 l/100 km at 100.7 k.p.h (Bordeaux/Paris : 600 km).

ECO 2000 (PSA) achieved 2.98 l/100 km with a spark ignition engine and 2.53 l/100 km world record with the Diesel version (standard EEC specific consumption).

VIRAGES is a programme aimed at reducing the fuel consumption and increasing the safety of heavy trucks : research is still underway.

Results from 10 years of research

These programmes led to heavy knock-on effects to French line production vehicles (Citroën AX, Peugeot 205 and 405, Renault Super 5 and 19,...).

The specific consumption of new French cars has dropped by 12 % over the 5-year period, from 1983 to 1988, and by 23 % since 1975, as already shown in figure 7.

More recent research programmes underway

Within the context of the 5-year overall research programme on land transport PRDTTT (Programme de Recherche, Développement Technologique sur les Transports Terrestres) initiated in 1982, a specific programme aimed at improving engine and combustion technology (Diesel and lean-burn engines) was launched by AFME and the Ministry of Research in 1986 ; emphasis was laid on organizing collaboration between car manufacturers and the basic research work carried out by university teams in the field of engine synthesis.

Results were encouraging but potential applications were broken up by the enforcement of new pollution standards especially NOx emissions ; a cleanish, very economical solution had to be ruled out in favour of 3-way catalyst exhaust systems. The withdrawal of the lean-burn solution is estimated to have resulted in a loss in saving of 1 MTOE/year by year 2000, plus an extra consumption of 1.5 MTOE/year due to the compulsory 3-way catalyst.

With consideration to the real car use patterns shown in studies such as EUREV, lean-burn and Diesel engine technology is probably better suited than 3-way catalysts for optimizing emission reduction, as the latter are not yet operational for short-distance cold-start journeys.

Alongside these programmes other research topics are also being investigated in the area of novel engine technology, the target being to optimize fuel consumption under low emission constraints e.g two-stroke engines and gas turbine engines (AGATA programme for EUREKA).

Within the second 5-year land transport research programme (PREDIT), the SERRE programme specifically adresses low consumption/low emission vehicles. The target of SERRE is to prevent the extra consumption of 1.5 MTOE/year expected by year 2000 with a business-as-usual scenario ; this very important project covers three time-scales :

- Short term : technology to be developed in order to halve the extra cost of anti-pollution devices without increasing specific consumption.

- Medium term : alternative solutions will be developed in order to strongly reduce consumption (and CO2 as a result).

- Long term : new research areas will be investigated with a mind to cutting out CO2 emissions (hydrogen combustion and fuel cells).

IV - BY WAY OF CONCLUSION

In the opinion of the panel, the situation prevailing in North America is very different from that of Europe and Japan.

Nevertheless, it is **absolutely essential** to take into account real use patterns when designing vehicles, particularly the engine components (high efficiency for cold-start short-mileage, high efficiency for transient conditions).

It is also essential to give a preference to technologies which produce good average behaviour over a wide spectrum of utilization, rather than high performance under stringent conditions.

Electric vehicles can be developed for dedicated urban fleets (low autonomy), hybrid electric/conventional propulsion vehicles for the common public.

Another field of action with important consequences on emissions and consumption is to improve traffic regulation technology and support policies aimed at reducing transient engine-running ; new road-use patterns could be implemented (parking policy, urban tolls, limited mileage,...) in order to make public transport more attractive.

Steps should be taken to improve driving ability, right from the learner stage : improved training for driving instructors, modern educational methods such as computer-assisted instruction and driving simulators,...

Engine tuning and maintenance should be improved, over and above safety considerations. Car repair mechanics should receive better training on this point, especially in Southern Europe.

Insofar as possible, policies should be worked out with a mind to ridding the roads of old and badly maintenanced vehicles.

Town planning has very important consequences on the energy consumption of the transport sector :

- rising distance from home to work

- towns increasingly planned with the assumption of private car ownership.

Reduction in polluting and CO2 emissions are expected to result mostly from traffic regulation, alternative town planning and the promotion of public transport.

From a long-term standpoint, strong action should be undertaken in the field of public education about transport considerations, and right from an early school age (what form of transport to use, how to drive properly,...).

ENERGY CONSUMPTION IN ROAD TRANSPORT IN FRANCE

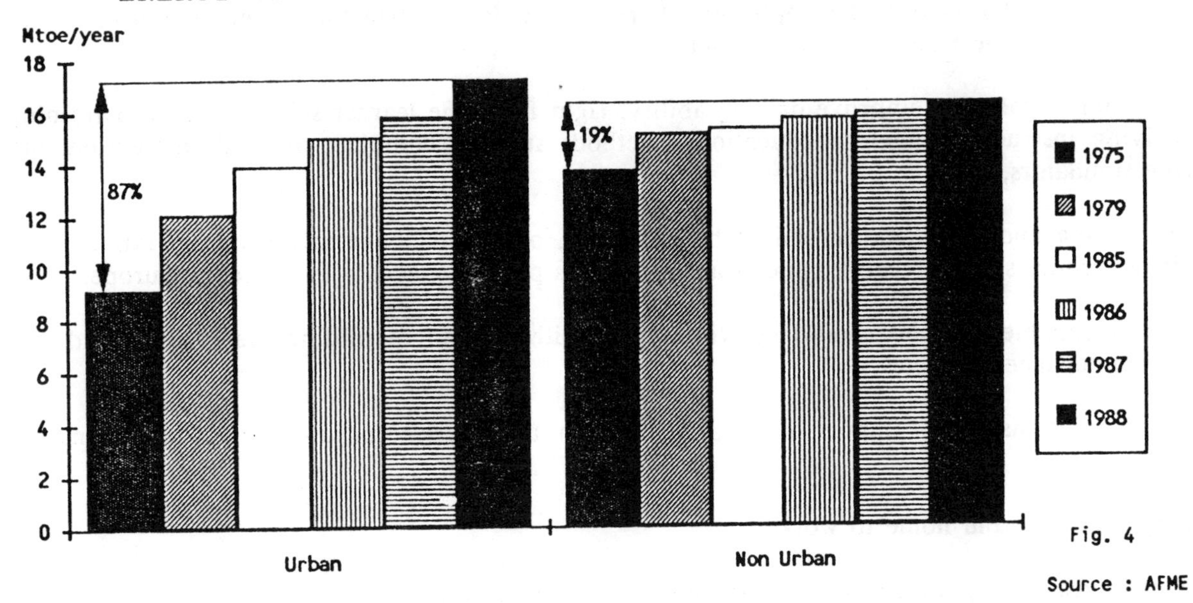

Fig. 4

Source : AFME

DISTRIBUTION OF JOURNEY LENGTHS

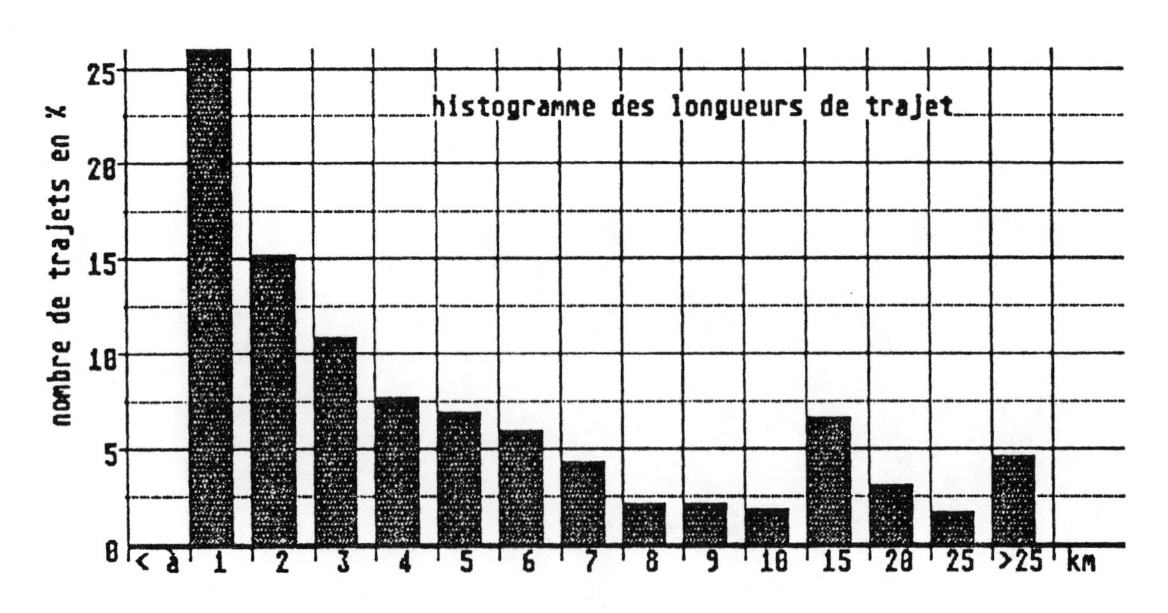

Fig. 5

Source INRETS

GROWTH IN OVERALL TRANSPORTATION DEMAND IN EUROPE

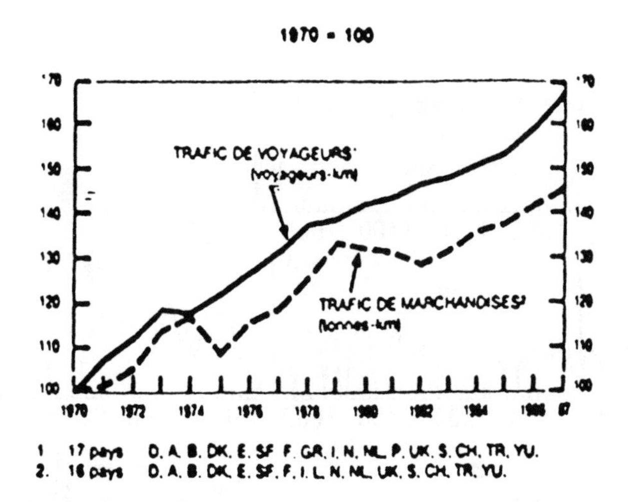

Fig. 1

PASSENGER TRANSPORTATION

FREIGHT TRANSPORTATION

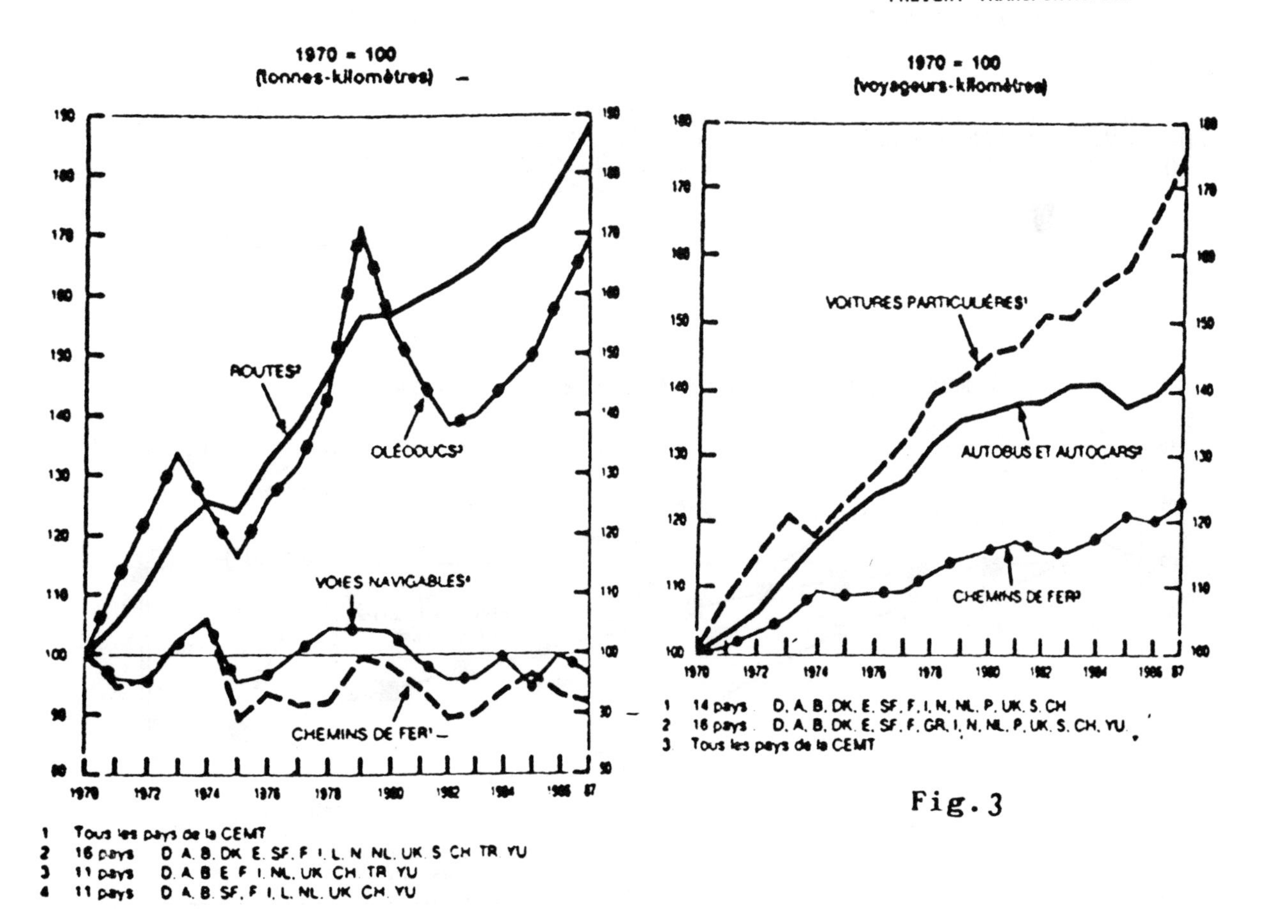

Fig. 3

Fig. 2

(Source CEMT)

FOUR CLASSES OF KINEMATIC CYCLES CHARACTERIZING REAL CAR USE PATTERNS - EUREV

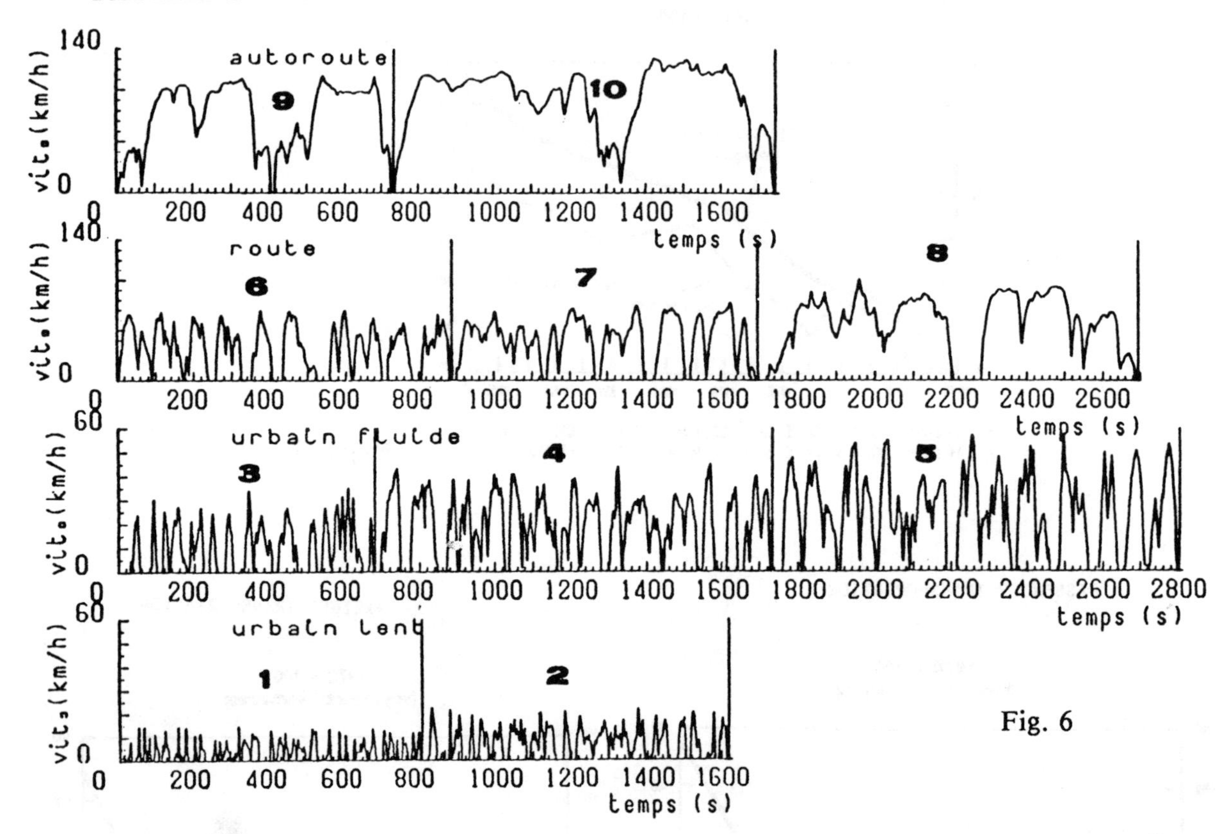

Fig. 6

Source INRETS

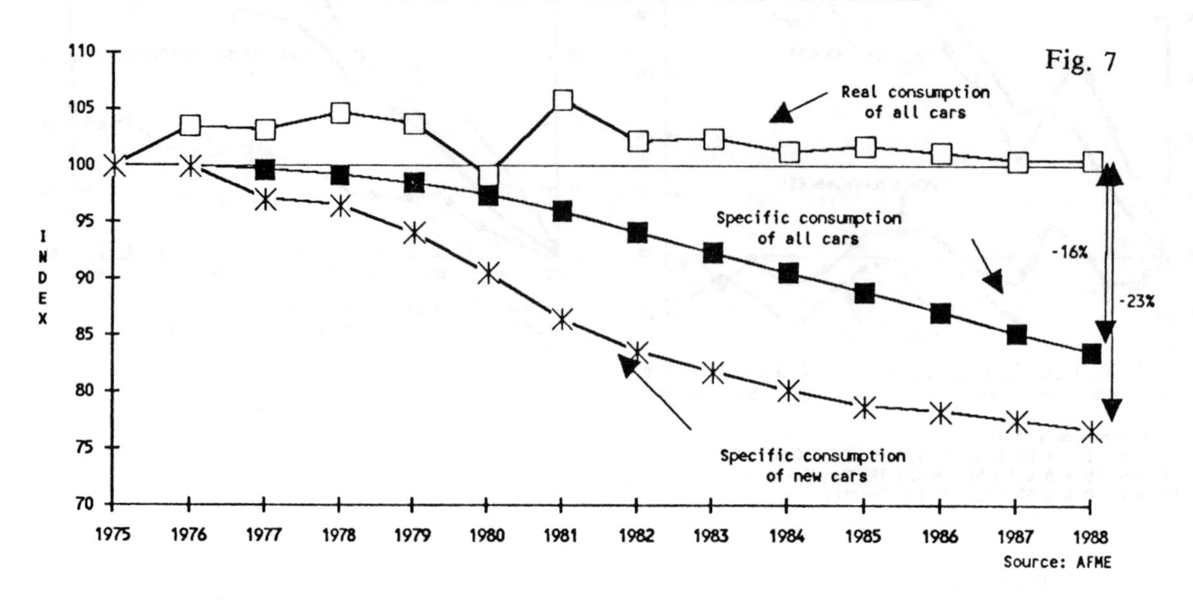

Cost Effectiveness of Future Fuel Economy Improvements

Carmen Difiglio, K.G. Duleep,** and David L. Greene****

U.S. petroleum use today is 2 million barrels per day lower than it would have been if automobile fuel economy had not improved since 1975. This paper explores the potential for and cost of further increases in domestic passenger car fuel economy using market-ready technologies and sales mix shifts. Using technology already included in manufacturers' production plans and based on consumers' willingness to pay for increased fuel economy, domestic auto mpg could be increased from the 1987 level of 27 mpg to 31.6 mpg in 1995 without reducing vehicle size or performance from 1987 levels. By 2000, 34.3 mpg can be justified on the same basis. A higher level, 36.4 mpg, is cost-effective, based on fuel cost savings over the entire expected vehicle life. The maximum level achievable with the technology included in this analysis is 39.4 mpg, but this level would not be cost-effective. Sales mix shifts stimulated by price subsidies for efficient cars and surcharges on inefficient models can cause about 1 or 2 mpg of higher fuel economy before becoming too costly.

INTRODUCTION

Improved automobile fuel economy has contributed and will continue to contribute to reducing U.S. dependence on imported oil. U.S. oil consumption today is 2 million barrels per day lower than consumption

Note: The views expressed in this paper are those of the authors and do not necessarily represent the views or policies of the U.S. Department of Energy, Energy and Environmental Analysis, Inc. or the Oak Ridge National Laboratory.

* U.S. Department of Energy
** Energy and Environmental Analysis, Inc.
*** Oak Ridge National Laboratory

would have been if automobile fuel efficiency had not improved since 1975.

The purpose of this paper is to provide estimates of the future level of fuel economy that can be achieved by the major U.S. domestic manufacturers in 1995 and 2000, on the basis of economic efficiency (cost effectiveness) and proven fuel economy technology. Although we present numbers to several significant digits throughout this paper, we recognize that our estimates are approximate. The extra detail is presented because our methods operate at this level and because it permits some checking of our results. We encourage readers to interpret our estimates as being in the neighborhood of the true mpg or cost levels.

METHODOLOGY

The analysis required to make these estimates begins with detailed data on fuel economy technologies. First, we identified proven fuel economy technologies already available in existing carlines or in prototypes. Next, we estimate the cost and efficiency improvement potential for each technology using information on manufacturers' production plans, engineering journals and trade publications, and comparative analyses of similar makes and models with and without the technology in question. Engineering analysis was used to verify the estimated efficiency improvement potentials. The result of this process is a database of market-ready technologies, their applicabilities to individual market segments, fuel economy improvement potentials, costs, and dates of availability (for details, see Energy and Environmental Analysis, Inc., 1986, 1987, 1988).

Numerous technologies with the potential to improve fuel economy are omitted from this analysis because either, 1) they are not market-ready, or 2) they do not presently meet vehicle emission standards, or 3) they detract significantly from performance, ride, or capacity, or in some other way are not acceptable to consumers.

Whether or not a specific technology will be used to improve the fuel economy of a particular carline depends on three factors: 1) applicability, 2) availability, and 3) cost-effectiveness. *Applicability* is an engineering judgment that considers technical compatibility of technologies with each other and with specific vehicle types, as well as whether the technology would degrade the acceleration, ride, or interior volume of the vehicle. In this analysis we do not use technologies in carlines where they would reduce performance, ride, or capacity over 1987 levels.

Availability is a question both of the readiness of the technology for marketing and the manufacturers' ability to produce it. Lead times of 3-5 years are required to redesign vehicles and construct or convert the necessary production lines. Furthermore, it is not reasonable to expect a manufacturer to redesign all his carlines at once since this would require an extraordinary level of investment, plus the scrappage of capital equipment with substantial useful life remaining. We present scenarios based on manufacturers' current plans to implement fuel economy technology but in one case we also assume maximum use of these technologies, regardless of the cost to consumers or the disruption of manufacturer plans. We intend the latter to provide an upper bound on what is reasonable.

Cost-effectiveness depends on the price the consumer must pay for the technology and the present value of the fuel cost savings it will produce over time. The estimated value of fuel savings reflects the consumer's discount rate, and expected vehicle use. We assume that gasoline costs $1.10 per gallon in 1995 and $1.32 in 2000, and that passenger cars have a 10-year lifetime. (Our price assumptions are consistent with the Energy Information Administration's 1989 Annual Energy Outlook, Base Case. 1987 $'s are used throughout this paper.)

Two approaches are used to discount future fuel cost savings. The one we believe most closely approximates consumer behavior and the way manufacturers will make their decisions assumes that car buyers discount fuel savings at a real rate of 10 percent per annum but only consider the first four years of the vehicle's life. This method implies that the market for mpg is imperfect. It is consistent with implicit discount rates in the range of 20 to 30 percent for savings over the full life of the vehicle found in other studies (Greene, 1986; Greene, 1983; Train, 1985). We call this the 4-year or "Product Plan" case because we believe it most closely represents the way manufacturers would plan the introduction of fuel economy technology. In the second method we assume a real discount rate of 10 percent over the full 10-year life of the vehicle. This yields a level of mpg that is cost-effective from a social viewpoint, since the economic welfare of all vehicle owners is accounted for, not just the first four years. For this reason we call this scenario the "Cost-Effective" case. However, it is not the true lowest social cost for two reasons. First, fuel prices include approximately $0.25 per gallon in road user fees. The fact that the car owner avoids paying these doesn't save society any money. Second, gasoline use generates externalities, including air pollution and traffic congestion, the cost of which is not included in the price. Unless the externalities and road user taxes are equal, our Cost-Effective case is not the true lowest social cost case.

Fuel economy technologies are selected for individual automobile types in specific years by an algorithm contained in the Department of Energy's (DOE) Technology Cost Segment Model (TCSM). The TCSM has been used by DOE over the last several years to provide technical comments to the Department of Transportation for their fuel economy rulemakings as well as for other energy policy analyses. In the TCSM, three factors determine a type of automobile, i.e., a carline:

- Manufacturer (GM, Ford or Chrysler)
- Market Segment (Subcompact, Compact, Intermediate, or Large/Luxury)
- Model year.

Each technology's applicability to each carline type has been evaluated.

The TCSM's algorithm works essentially as follows: Technologies are ranked according to the ratio of their initial cost to the percent mpg improvement they produce. The technology with the lowest cost to improvement ratio is applied first, then the one with the next lowest, and so on, until a technology is reached that is not cost effective at the assumed price of gasoline and the current mpg level. Calculations of fuel economy by vehicle type are subject to engineering controls that reflect the way technologies interact with each other and account for production constraints that limit the availability of technologies to quantities that are not too small to be economical.

The TCSM has been continuously updated based on domestic manufacturers' fuel economy plans submitted to the Department of Transportation as well as other sources, to reflect the latest technical information on the costs and fuel economy benefits of specific fuel economy technologies (Energy and Environmental Analysis, Inc., 1981, 1985 and 1986).

The fuel economy improvement achieved by a set of available, applicable, and cost-effective technologies is the sum of their individual percent mpg improvements. The individual mpg improvements are added to calculate the total percent improvement in mpg. It can easily be seen that one plus the sum of percent mpg improvements, P_i, is less than or equal to the product of the one plus P_i's.

$$(1 + \sum_i P_i)\, mpg \leq mpg\, \pi (1 + P_i) \quad (1)$$

The righthand side of equation (1) includes a number of extra cross-product terms, all of which are positive. The costs of technologies used are likewise summed to obtain the cost of the mpg improvement.

New car fuel economy can also be increased by shifting sales away from less efficient carlines and configurations toward more efficient ones using price incentives and disincentives. However, such deviations from the market-determined sales mix will cause an economic loss. In general, the total loss will be divided between lower manufacturer profits and lost consumers' surplus. We have developed a method of approximating the size of this loss by assuming manufacturers bear the full economic cost. Holding expected consumers' satisfaction constant, we estimate the price increase or decrease for each carline that would be required to achieve an mpg target. We do this by minimizing an objective function equal to the weighted sum of the squared change in consumers' surplus caused by the price changes plus the squared deviation of salesweighted average mpg from the target level of fuel economy (the measure of consumers' surplus used is explained in the Appendix). A solution occurs when a set of prices is found for which the objective function equals zero. The salesweighted average of these price changes constitutes the average subsidy, or cost, per car of achieving higher fuel economy. A detailed exposition of the technique can be found in Greene (1989). We assume these shifts in market share occur only after all technological fuel economy improvements are made.

As more efficient new cars are added to the existing stock of vehicles, the total vehicle fleet mpg will gradually improve. A third model, the DOE Motor Fuel Consumption Model (MFCM), is used to estimate the impacts of changing the fuel economy of new vehicles on total gasoline and diesel fuel consumption (Energy and Environmental Analysis, Inc., 1987). The MFCM is an accounting model. Estimates of new car sales are supplied by the model user. The MFCM uses historical average rates of vehicle use and vehicle scrappage, by age of car, to update the stock of vehicles, estimate miles traveled by vintage and, using vintage-specific fuel economies, compute fuel use. For each vintage, fuel use equals the number of vehicles times the average annual miles for vehicles of that age, divided by the average miles per gallon for the model year in question. The MFCM discounts estimated new car mpg by approximately 15 percent (the discount varies by model year) to account for the shortfall between EPA-tested new-car fuel economy and actual on-road fuel economy.

CURRENT SITUATION

Average U.S. new car fuel economy has been steadily increasing (Figure 1). This increase is due to the dramatic fuel economy improvement of domestically produced automobiles. The fuel economy of imported automobiles has not increased as rapidly, and has actually declined since 1983. This is due in part to the increased market share of large and midsize imported automobiles and also to changes in their performance, level of luxury accessories, and other characteristics.

Figure 2 illustrates trends in key vehicle characteristics related to fuel economy. It is apparent that the interior space of vehicles has remained more or less constant after a slight decline from 1979 to 1980.

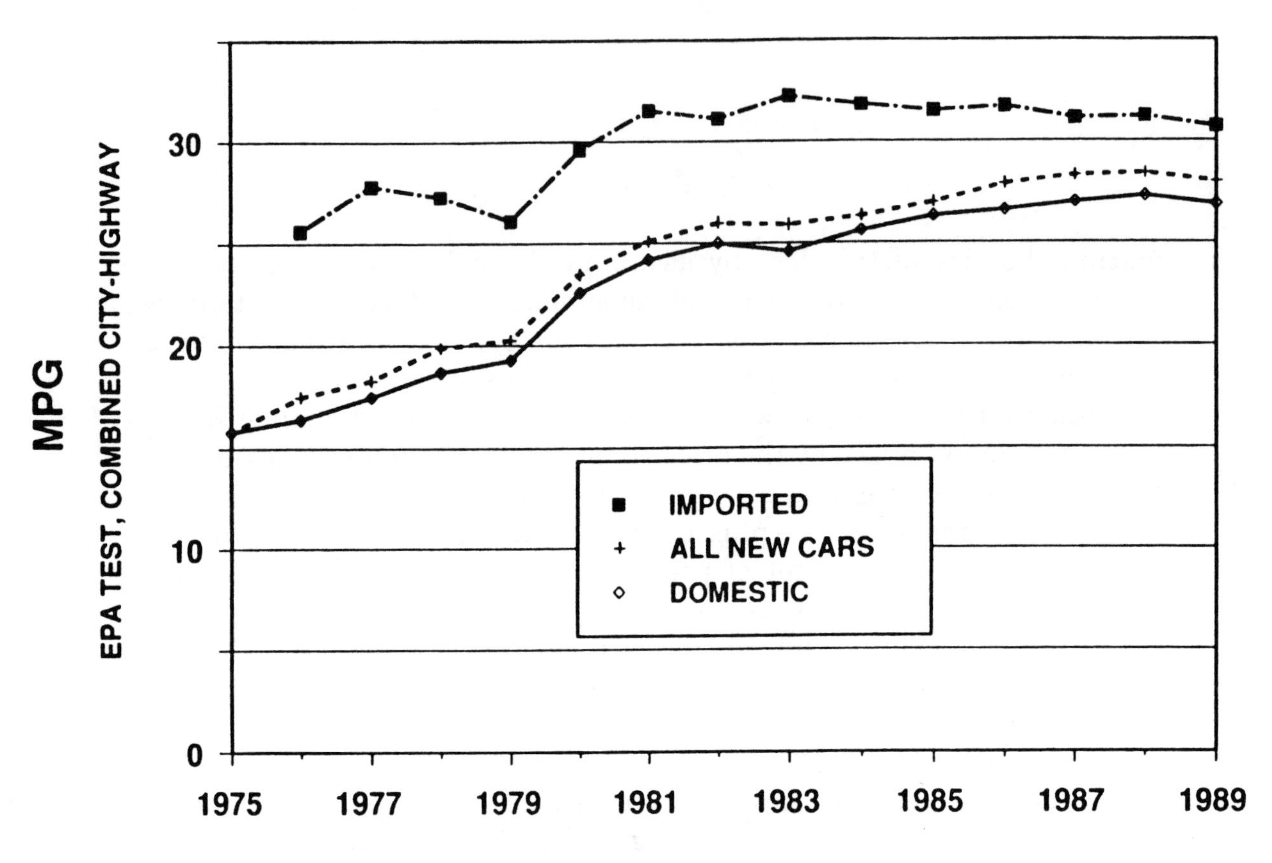

Figure 1. Trends in New Passenger Car Fuel Economy, 1975-1989

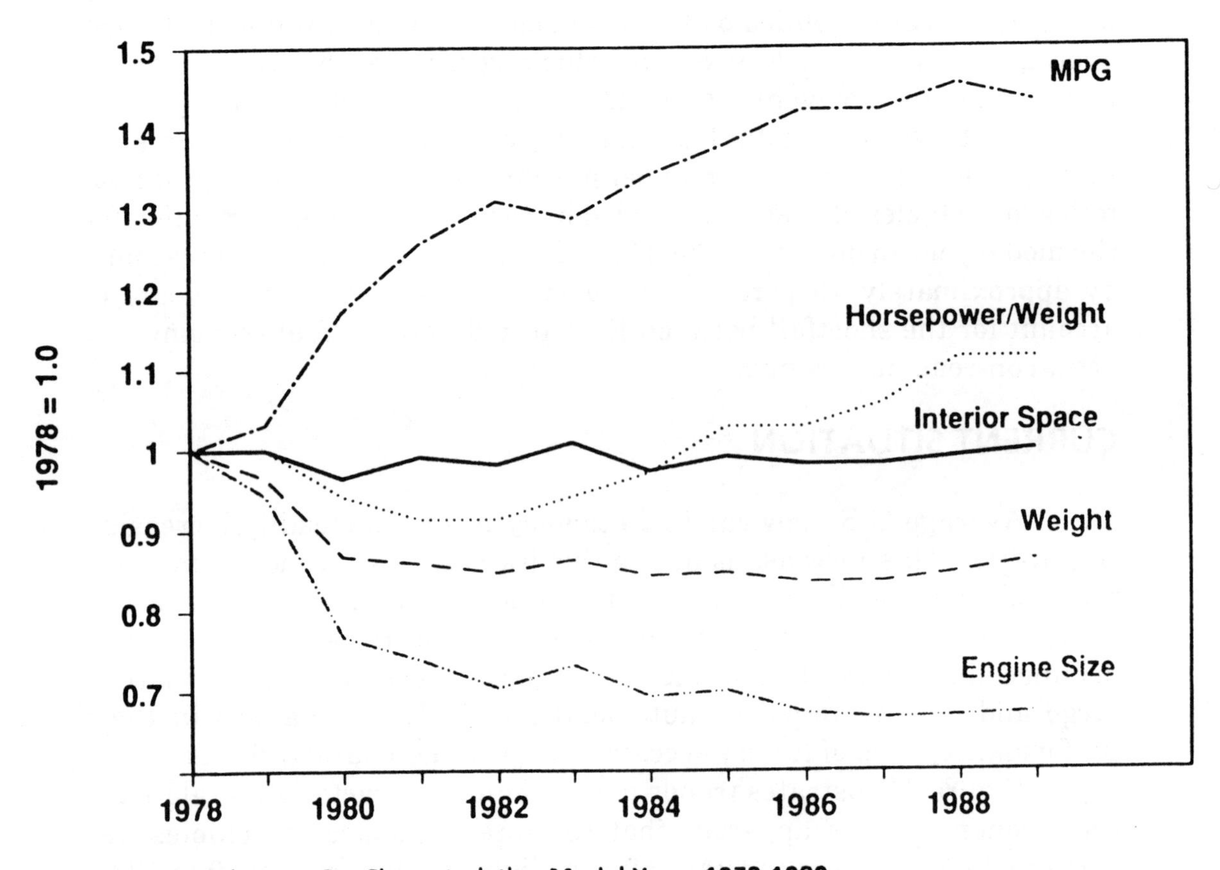

Figure 2. Trends in New Car Characteristics, Model Years 1978-1989

However, exterior car size, curb weight, and engine size have all declined significantly. This reflects the space efficiency of the newer front-wheel drive automobiles over the previous rear-wheel drive models, reductions in vehicle weight from front-wheel drive configuration and other factors, the ability of smaller engines to power the lighter automobiles, and increased horsepower from smaller engines. Indeed, performance measured by acceleration time or horsepower-to-weight ratio has never been higher than at present (Heavenrich and Murrell, 1989). While significantly more fuel efficient, current automobiles offer interior accommodations equal to past autos and provide levels of acceleration, braking and roadholding that surpass past levels. The current situation reflects a high degree of technical progress by the international automobile industry to improve automobile fuel economy while also increasing the consumer utility of automobiles compared to previous model years.

Fuel efficiency, as measured by Corporate Average Fuel Economy (CAFE), for domestic automobiles was 26.9 mpg in model year (MY) 1987, nearly twice what it was in 1974. This fuel economy progress has been vital to reducing U.S. oil imports. U.S. oil consumption today is 2 million barrels per day lower than consumption would have been if automobile fuel efficiency had not improved since 1975.

It is important to keep in mind that current automobiles are expensive in comparison to the annual fuel costs they incur. A typical, new, domestic, 1988 model-year automobile cost $14,500 (1987 $), yet only consumes 370 gallons of fuel per 10,000 miles traveled (about the average number of miles driven per year for the U.S. fleet). At $1.10 per gallon, this typical annual fuel expense represents about 3 percent of the purchase cost. In addition, insurance plus maintenance expenses typically constitute a larger component of operating costs than fuel expenses. When capital cost is also included, total lifecycle fuel expenses end up being only about 15 percent of the total cost of owning and operating an automobile (Davis, Shonka, and Hu, 1988).

These facts are highlighted in order to put the analytic results that follow into perspective. The desire to reduce fuel expenses, even with higher gasoline prices, will not motivate consumers to significantly alter their preferences for makes and models of automobiles. However, higher fuel expenses *will* justify greater use of fuel economy technologies by manufacturers. Once all fuel economy technologies have been used to improve fuel economy, the only way to further improve fuel economy is for consumers to purchase different cars, perhaps lighter, smaller or with less performance or luxury features than they would prefer. Since fuel expenses are currently such a small proportion of the cost of owning and operating a car, fuel prices would have to be very high indeed to cause this to happen.

Table 1. Potential Fuel Economy of Domestic Automobiles in 2000

Technology	% MPG Benefit 1995 vs. 1987	% MPG Benefit 2000 vs. 1995	1987 Market Share	Product Plan (%) 1995 Share	Product Plan (%) MPG Gain	Product Plan (%) 2000 Share	Product Plan (%) MPG Gain	Cost-Effective (%) 2000 Share	Cost-Effective (%) MPG Gain	Max Technology (%) 2000 Share	Max Technology (%) MPG Gain
Engine Improvements											
Intake Valve Control	10.0	10.0	0%	0	0	20	1.20	50	3.00	70	4.20
Overhead Cam Engine	6.0	6.0	24%	69	2.70	99	1.80	99	1.80	99	1.80
Roller Cam Followers	1.5		55%	95	0.60	95		95		95	
Low-friction Pistons/Rings	2.0		20%	100	1.60	100		100		100	
Adv. Friction Reduction		2.0	0%	0		80	1.60	80	1.60	80	1.60
Throttle-Body Fuel Injection	3.0		28%	40	0.36						
Multi-point Fuel Injection	3.0	3.0	48%	60	0.36	100	1.20	100	1.20	100	1.20
4-Valve 6 cyl. for 8 cylinder	10.0	10.0	0%	12	1.20	16	0.40	20	0.80	20	0.80
4-Valve 4 cyl. for 6 cylinder	10.0	10.0	0%	18	1.80	24	0.60	30	1.20	30	1.20
4 Valve 4 cyl. for 4 cylinder		5.0	0%	0		10	0.50	40	2.00	50	2.50
Transmission Improvements											
Electronic Control	1.5		0%	80	1.20						
Torque-converter Lock-up	3.0		60%	80	0.60						
4-Speed Automatic	4.5		40%	80	1.80						
5-Speed Automatic		2.5	0%	0		10	0.25	20	0.50	40	1.00
Continuously Variable		2.5	0%	0		10	0.25	40	1.00	40	1.00
Other Improvements											
Front Wheel Drive	10.0	10.0	74%	86	1.20	95	0.50	99	1.30	99	1.30
Weight Reduction (Materials)		6.6	0%	0		0		0		80	5.28
Aerodynamic Drag Reduction I	2.3		20%	100	1.84						
Aerodynamic Drag Reduction II		2.3	0%	0		10	0.23	10	0.23	80	1.84
Electric Power Steering		1.0	0%	0		5	0.05	5	0.05	60	0.60
Lubricants/Tires	1.0		0%	100	1.00						
Tires		0.5	0%	0		20	0.10	100	0.50	100	0.50
Accessories	1.0	1.0	20%	100	0.80						
Total Fuel Economy Increase (%) Over 1987 Base for 1995, over 1995 for 2000 scenarios.					17.06		8.68		15.18		24.82
Estimated CAFE Based on 1987 = 27.0 mpg					31.6		34.3		36.4		39.5
Increased Consumer Cost (over prior case)					$240		$133		$120		$264
Increased Consumer Cost (over 27 mpg)					$240		$373		$493		$757

COSTS OF HIGHER FUEL ECONOMY

We present estimates of Product Plan fuel economy for 1995 and Product Plan and Cost-Effective fuel economy for 2000. We also estimate a "Maximum Technology" 2000 level and higher levels which require sales mix shifts. All represent changes from 1987 levels, assuming no change in the size, performance, or ride quality of 1987 MY cars. In reality, acceleration rates and vehicle size have increased since 1987, so that our projections assume lower levels of performance and size than current (1989) and planned 1995 levels. Table 1 lists the fuel economy technologies included in this analysis. All are proven technologies that are either in production on certain carlines or included in manufacturers' plans for future product lines. We have excluded certain important technologies either because they are unproven, degrade vehicle utility, do not meet emissions standards, or have proven to be unsuccessful in the market in the past. Most prominent among the omitted technologies is greater use of diesel engines. Environmental regulations may make future use of this technology infeasible. However, further technological advances or much higher fuel prices could lead to market acceptance of additional fuel economy technologies. In that case, higher levels of new car fuel economy would be possible.

Given 1987 levels of vehicle performance, existing manufacturer plans for introducing new carlines and fuel economy technology, and gasoline at $1.10 per gallon, 1995 new car fuel economy would be 31.6 mpg. This Product Plan case is consistent with manufacturers' planned use of fuel economy technology for 1995. However, it does not necessarily correspond to MY 1995 vehicle size, luxury, or performance, since we hold these attributes constant at MY 1987 levels. It achieves a 4.6 mpg improvement over the Base case (1987 domestic new car fuel economy = 27.0 mpg). The technological improvements which cause this 4.6 mpg increase are listed in Table 1 with the estimated percentage fuel economy gain and increased penetration into the domestic fleet for each technology. The increase in cost to the consumer compared to the Base case is estimated to be $240.

The Product Plan case for MY 2000 results in an average domestic fuel economy of 34.3 mpg and adds $133 of increased consumer cost over the 1995 Product Plan case. The Cost-Effective case (MY 2000) accounts for discounted fuel savings over the full 10-year expected lifetime of the car. This case results in an average domestic fuel economy of 36.4 mpg and adds $120 over the MY 2000 Product Plan case.

The maximum feasible fuel economy in MY 2000 using technology alone (without changing car sizes or other consumer attributes) is 39.4

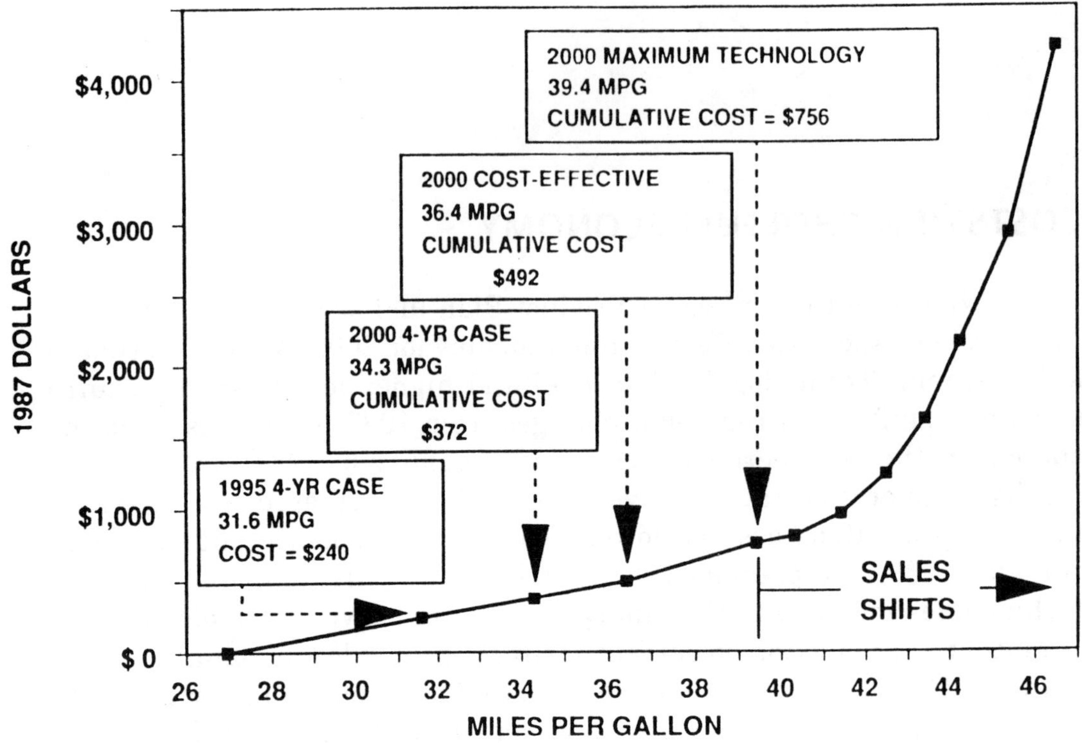

* Increase over 1987 baseline of 27 MPG for domestic manufacturers.

Figure 3. Cumulative Cost of 19.4 MPG Increase in New Passenger Car Fuel Economy by 2000

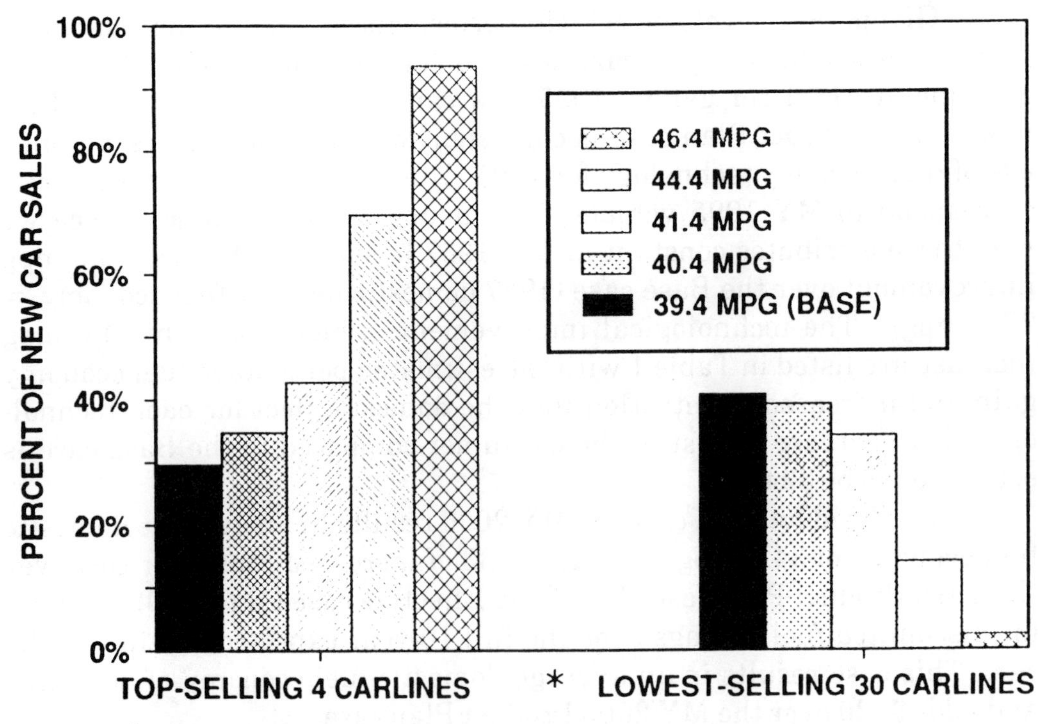

* Which carlines are highest and lowest selling changes
as sales are shifted to achieve higher MPG.

Figure 4. Increasing MPG Via Sales Shifts Concentrates Sales in a Few Carlines

mpg (Maximum Technology case). The increased cost per vehicle compared to the 1995, 31.6 mpg case is estimated to be $516, for a cumulative cost of $756 over the 1987, 27.0 mpg baseline. In the Maximum Technology case, every market-ready technology is used on every carline for which is makes engineering sense. In addition to the technologies used in the prior cases, 5-speed automatic transmissions, further reductions in aerodynamic drag, weight reduction through use of advanced materials, intake valve control, and the use of smaller 4-valve per cylinder engines to replace 4-cylinder engines are used, resulting in MY 2000 fuel economy of 39.4 mpg. The summary of technologies, their market penetration, and estimated benefit are shown in Table 1.

Manufacturers could achieve a higher level of fuel economy by 1995 by accelerating the introduction of some of the technologies we have added for MY 2000. However, they would probably not do this under normal business conditions. For example, 4-valve engines could replace 100 percent of current engines. However, such a precipitous turnover of production facilities in such a short time would cost considerably more than we have estimated. If manufacturers were to strive for 39.4 mpg in 2000, however, they would have to accelerate some of these changes and reach a higher level of fuel economy in 1995, at a higher cost, than the 31.6 mpg of our Product Plan case.

After use of all fuel economy technologies and some degree of performance reduction, the primary way to further increase fuel economy is to alter the mix of vehicles--more efficient cars--fewer inefficient cars. The cost of such a shift is estimated using the constant consumers' surplus methodology described above and in the Appendix. For each mpg increase beyond the 39.4 mpg Maximum Technology case, we estimate the average subsidy per car that would be required to achieve the desired average mpg while holding consumers' satisfaction constant. The required subsidy measures the cost of the mpg increase. Our estimates are approximate due to the necessary assumptions and limitations of our method and should not be considered precise to the last dollar. They do, however, generally indicate the magnitude of the cost of manipulating the sales mix for higher mpg.

The TCSM has forty carline-engine transmission combinations representing the offerings of the three major domestic manufacturers. We estimated an initial sales distribution for 1995 using the 1987 sales distribution and, for new carlines, the sales of the obsolete models they replace.

A 1 mpg improvement to 40.4 mpg requires an average subsidy of $50 per car. This subsidy is roughly equivalent to the present value of discounted fuel cost savings over a 10-year lifetime at $1.10/gal. It is also, by coincidence, the same as the statutory penalty for falling short of the mandated CAFE standard ($5 per 0.1 mpg, per car). A two mpg increase costs $207 per car, equivalent to over $2/gal., similarly discounted. The cost of still higher fuel economy increases at an increasing rate, as Figure 3 illustrates. The sales mix shifts also become drastic. At 44.4 mpg the single highest mpg carline must capture almost one third of total sales. At 46.4 mpg, the limit of what can be achieved with sales mix shifts, the same carline constitutes 70 percent of all car sales. Thirty-two carlines each capture 0.3 percent or less of the market, which means that most would probably be discontinued due to uneconomically low production volumes. This would leave only eight of the original 40 carlines in production. The social cost of this radical alteration is estimated at about $3,500 per car, or $24 billion per year (for a sales volume of 7 million domestic new cars).

Figure 4 illustrates how sales become increasingly concentrated in a few high-mileage product lines as mpg is pushed to the limit. In reality, low selling carlines will be discontinued when sales fall below economic production levels. As a result, consumer choice would be greatly restricted.

We do not propose that such a change would ever occur. Also, the model's parameters would not remain valid for such a drastic shift of automobile purchases. Buyers of inefficient luxury cars, for example, may be relatively insensitive to price increases and, therefore, require significantly larger inducements than our parameters imply to purchase more efficient models. Therefore, we believe that our method underestimates the social cost of such large mpg increaes.

This analysis of consumer surplus is presented with the full realization that we are dealing with theoretically based estimates. However, these estimates would have to be in substantial error before their implications were reversed.

Finally, these estimates of economic loss must be considered short-run effects only. Given time, manufacturers would redesign vehicles or develop new technology rather than face the high cost of increases of 2 mpg and higher. What the analysis does clearly demonstrate is that sales mix shifts are a very expensive way to push fuel economy more than 1 to 2 mpg beyond what can be achieved by using fuel economy technologies.

COST-EFFECTIVENESS AND PETROLEUM SAVINGS

A summary of the cost effectiveness of each level of fuel economy improvement is provided in Table 2 and Figure 3. Column 1 of Table 2 gives the gasoline price per gallon used to decide the cost effectiveness of the marginal fuel economy technology. Columns 5 and 7 show the aver-

Table 2. Cost Effectiveness of Different Levels of Fuel Economy

	Marginal Cost = Savings $/gal.	*EPA mpg*	*On-Road mpg*	*Cost Increase*	*Average Cost per Unit of Fuel Saved Relative to Prior Case*			
					4-year $/gal	***4-year $/bbl***	***10-year $/gal***	***10-year $/bbl***
Base Case (1987 Fuel Economy)	NA	27.0	23.0	NA	NA	NA	NA	NA
Product Plan 1995 (4-yr., 10% discount)	$1.10	31.6	26.9	$240	$0.93	$15	$0.55	$1
Product Plan 2000 (4-yr., 10% discount)	$1.32	34.3	29.2	$133	$1.11	$22	$0.66	$5
Cost-Effective 2000 (10-yr., 10% discount)	$1.32	36.4	30.9	$120	$1.48	$36	$0.88	$13
Maximum Technology (Constant Utility)	NA	39.4	33.5	$264	$2.62	$78	$1.55	$38
Mix Shift (First 2 mpg)	NA	41.4	35.3	$207	$3.52	$111	$2.09	$58
Mix Shift (Last 5 mpg)	NA	46.4	39.4	$3,138	$25.07	$907	$14.87	$530

The 4-year discounting is based on the discounted value of 4 years of automobile travel (40,865 miles). The equivalent number of miles for the 10-year calculation is 68,900.

age cost per gallon of fuel saved (based on a 15 percent discount of EPA test fuel economy to adjust for on-road mpg shortfall) for the increased use of all technologies included in each scenario. The 1995 Product Plan case produces an average fuel economy of 31.6 mpg in MY 1995. This level is cost-effective at the margin, assuming consumers account for 4 years of fuel consumption with an implicit discount rate of 10 percent. The entire 4.6 mpg *increment* of fuel economy improvement would be paid for by the resulting fuel savings if gasoline cost $0.93 per gallon (or oil cost $15 per barrel). It should be noted that this case includes a technology that is cost-effective at $1.19/gal. However, this technology, the 4-speed automatic transmission, provides significant performance benefits that justify an expense over and above that offset by fuel cost savings.

In the year 2000, the level of mpg in the Product Plan (4-year fuel savings) case is 34.3 mpg. This Product Plan case results in higher fuel economy than in 1995 because fuel prices are assumed to be higher ($1.32 instead of $1.10 in 1995) and additional technologies are estimated to be available in 2000 that were not available in 1995. The Cost-Effective case is only estimated for the year 2000 and accounts for discounted fuel savings over the full 10 years expected lifetime of the car. This case results in an average domestic fuel economy of 36.4 mpg and adds $120 over the 2000 Product Plan case. We do not present a Cost-Effective case for the year 1995 because it would call for too many changes to existing product lines to be achievable by 1995 without significantly increasing the consumer cost above our estimates which assume a normal turnover of production equipment. We have not estimated the extra costs of accelerated capital turnover that would be required to go beyond the Product Plan case for 1995.

The average cost per gallon of gasoline saved is a useful measure of cost effectiveness and highlights the potential importance of market imperfections in automotive fuel economy. For example, to justify the cost increment of $120 for 36.4 mpg versus 34.3 mpg the fuel saved would have to be valued at $1.48 per gallon ($36 per barrel of oil) if discounted over only four years. If one discounts the savings over the full vehicle life of 10 years, however, the average cost per gallon of gasoline saved falls to $0.88/gal. This is well below the fuel price at which the marginal cost just equals the marginal savings accrued over 10 years and discounted at 10 percent ($1.32/gal.).

The maximum fuel economy achievable without changing market mix or performance in MY 2000 is 39.4 mpg. The increased cost per vehicle over the 36.4 mpg case is estimated at $264. Gasoline would have to cost $2.62 per gallon ($78 per barrel of oil) to repay this cost (4 years, 10 percent discount rate). Discounted over 10 years the cost per gallon of gasoline saved is $1.55 ($38/bbl.), still $0.23 higher than the projected market price.

The last two cases presented in Table 2 assume mix shifts among vehicles with the fuel economies achieved in the maximum technology case. The first of these mix shift cases increases average fuel economy from 39.4 to 41.4 mpg. This 2 mpg increase results in a loss of $207 that would only be justified if fuel cost $3.52/gallon (4-year fuel savings) or $2.09/gallon (10-year savings). An additional 5 mpg improvement in average fuel economy is theoretically possible. This extreme 46.4 mpg case results in such a large estimated surplus loss that fuel prices would have to be over $25/gallon (4-year savings) or $14/gallon (10-year savings). Measured in terms of cost per barrel, the extreme 46.4 mpg case would require oil prices in the range of $500-$900/bbl. to be cost-effective. As pointed out above, manufacturers would develop new technologies rather than attempt to achieve such drastic changes in sales mix. This case is presented only to illustrate how sales mix shifts have a relatively limited potential to improve fuel economy.

It is also worth noting that the estimated price subsidy required for the first mpg is only $50. This level is roughly cost effective with gasoline at $1.10 (again assuming 4-year, 10 percent discounting of fuel savings). This result hints that manufacturers may already be using pricing strategies to achieve small (1 mpg or less) mpg gains in order to meet CAFE requirements.

A summary of oil savings resulting from different fuel economy levels is presented in Table 3. The Base case assumes new car fuel economy remains constant at 27 mpg through the year 2010. Every other case increases fuel economy to 31.6 mpg in 1995, and then to one of four higher levels in 2000. The Product Plan case (34.3 mpg in 2000) results in 790 thousand barrels per day of oil savings in 2005, increasing to 930 thousand barrels by 2010. The Cost-Effective case (36.4 mpg in 2000) would add another 130 thousand barrels per day in 2005 for a total savings of 920 thousand barrels per day in 2005. By 2010, the Cost-Effective case would save 1.1 million barrels per day over the Base case.

The Maximum Technology case (39.4 mpg) would result in 1.1 million barrels per day less oil use in 2005 and 1.35 million in 2010 relative to the Base case. The extremely expensive 46.4 mpg case adds relatively little in fuel savings--330 thousand barrels per day in 2005 and 420 thousand in 2010. This is because the difference in fuel consumption between a car that gets 39.4 mpg and one that gets 46.4 mpg is quite small--only 38 gallons for 10,000 miles of travel.

All of the quantities saved over the Base case are still increasing in 2010, for two reasons. First, although the average life expectancy of an automobile is 10 years, half survive longer than ten years so that

Table 3. Motor Fuel Consumption of Different Levels of Fuel Economy

	Fuel Economy (mpg)	Automobile Fuel Use in 2005/2010 (MMB/D) Total	Savings (Prior Case)	Accumulated Savings
Base Case (1987 Fuel Economy)	26.9	5.32/5.64	NA	NA
2000 Product Plan (4-yr., 10% Disc.)	34.3	4.53/4.71	0.79/0.93	0.79/0.93
Cost-Effective 2000 (10-yr., 10% Disc.)	36.4	4.40/4.54	0.13/0.17	0.92/1.10
Maximum Tech Case (Constant Utility)	39.4	4.22/4.29	0.18/0.25	1.10/1.35
Mix Shift (Maximum)	46.4	3.89/3.87	0.33/0.42	1.43/1.77

there are a sizeable number of MY 2000 or older cars still to be replaced in 2010. Second, vehicle miles of travel continue to grow beyond 2010, thereby increasing the fuel saved per fleet average mpg.

If we restrict ourselves to considering improvements that can be brought about by technology--not changes in consumer choice, the Product Plan case alone will result in 69 percent of the fuel savings that are achievable by 2010 with the technology listed in Table 1. Adding the further improvements of the Cost-Effective case would raise this to 81 percent. While the remaining 250 thousand barrels per day of oil savings would be desirable, these savings could not be achieved unless automobile purchasers accepted increases in vehicle prices that could not be justified by the value of the fuel savings they would realize. Energy security or environmental benefits might justify all or part of these increased consumer costs, but it is not the purpose of this paper to make that evaluation. These estimates of the cost of increased automobile fuel savings can be used, with estimates of other ways to reduce oil demand, to devise a least-cost approach to reducing U.S. oil consumption.

APPENDIX

A useful measure of consumers' surplus in situations where consumers choose among discrete alternatives (such as carlines) was devised by Williams (1977) and placed in the context of economic theory by Small and Rosen (1981). It is based on the multinomial logit choice model (see McFadden, 1973), in which the probability of choosing carline i, and hence its market share in the aggregate, is given by,

$$P_i = \exp\{U_i\}/\mathrm{Sum}_j[\exp\{U_j\}].$$

U_i is the utility of alternative (carline) i, and is assumed to be a function of the price change for carline i, plus a constant.

$$U_i = a_i + B(\mathrm{PRICE})_i.$$

The constant, a_i, represents all other characteristics of the carline that account for its utility to the consumer and, thus, its success in the market.

Williams (1977) and Small and Rosen (1981) showed that, for the multinomial logit choice model, when characteristics of the choices change, the change in consumers' surplus (CS) is,

$$CS = -(1/B)[\ln\{\mathrm{Sum}_j[\exp\{U_j'\}]/\mathrm{Sum}_k[\exp\{U_j\}]\}],$$

where U_j' indicates the utility of alternative j after the change (in this case a change in price). To compute CS for a change in prices, we need values for B and the a_i.

The a_i can be easily computed from a set of initial sales by carline. If P_{io} is the initial market share of vehicle type i and we arbitrarily set $\mathrm{Sum}_i a_i = 0$, then the a_is can be computed from the observed (current) market share as follows.

$$a_i = \ln\{P_i\} - (1/N)\,\mathrm{Sum}_i \ln\{P_i\}$$
$$a_i = \ln\{P_i/P_i\} + a_i$$

If we had not set $\mathrm{Sum}_i a_i = 0$, there would have been an infinite number of solutions for the set of a_i's. The constraint $\mathrm{Sum}_i a_i = 0$ determines a unique solution for the a's. The value of (CS) is independent of the arbitrary choice of a constraint to obtain a unique solution for the a_is.

Thus, given an initial set of market shares, the only parameter we need is B. After surveying the automobile choice model literature,

Greene and Liu (1988) suggested that B = -0.00056 is a midpoint estimate for automobile choice models published to date. We use that value here.

REFERENCES

Cardell, N.S. and F.C. Dunbar. (1980). "Measuring the Societal Impacts of Automobile Downsizing," *Transportation Research.* 14A:5-6.

Davis, S.C., D.B. Shonka, and P.S. Hu (1988). *1988 Transportation Energy Data Book*, manuscript, Oak Ridge National Laboratory, Oak Ridge, TN, May 31.

Energy and Environmental Analysis, Inc. (1981). *The Technology Cost Segment Model Post-1985 Automotive Fuel Economy Analysis, Final Report.* Prepared for the U.S. Department of Energy under Contract Number DE-AC01-79PE-70032.

Energy and Environmental Analysis, Inc. (1985). *Technology and Cost Data for Automobile Efficiency Analysis.* Prepared for Martin Marietta Energy Systems, Inc. (Oak Ridge National Laboratory).

Energy and Environmental Analysis, Inc. (1986). *Analysis of Current and Projected Capabilities of Domestic Automobile Manufacturers to Improve Corporate Average Fuel Economy.* Prepared for the U.S. Department of Energy under Contract Number DE-AC06-76 RL0183.

Energy and Environmental Analysis, Inc. (1987). *The Motor Fuel Consumption Model, Thirteenth Periodical Report.* Prepared for Martin Marietta Energy Systems, Inc. (Oak Ridge National Laboratory).

Energy and Environmental Analysis, Inc. (1988). *Light Duty Truck Fuel Economy, Review and Projections 1980-1995*, DOE/OR/21400-HO, prepared for U.S. Department of Energy under Contract No. DE-AC05-84OR-21400, Washington, D.C.

Greene, D.L. (1983). "A Note on Implicit Consumer Discounting of Automobile Fuel Economy: Reviewing the Available Evidence," *Transportation Research* 17B(6): pp. 491-499.

Greene, D.L. (1986). "The Market Share of Diesel Cars in the U.S., 1979-83," *Energy Economics* 8(1): pp. 13-21.

Greene, D.L. (1989). "The Cost of Short-Run Pricing Strategies to Increase Corporate Average Fuel Economy," unpublished, Office of Policy Integration, Office of Policy, Planning, and Analysis, U.S. Department of Energy, Washington, D.C.

Greene, D.L. and J. Liu. (1989). "Automotive Fuel Economy Improvements and Consumers' Surplus," *Transportation Research* 22A(3): pp. 203-218.

Heavenrich, R.M. and J.D. Murrell (1989). *Light Duty Automotive Technology and Fuel Economy Trends Through 1989*, EPA/AA/CTAB/89-04, U.S. Environmental Protection Agency, Ann Arbor, MI.

Hu, P.S. and L.S. Buckels. (1987). *Motor Vehicle MPG and Market Shares Report: Model Year 1986.* Oak Ridge National Laboratory, ORNL-6351.

Hu, P.S. (1988). "Motor Vehicle MPG and Market Shares Report: First Quarter of Model Year 1988," prepared for the U.S. Department of Energy under Contract Number DE-AC05-84OR21400.

McFadden, D. (1973). "Conditional Logit Analysis of Qualitative Choice Behavior," in *Frontiers of Ecuonometrics*". P. Zarembka, ed. New York: Academic Press.

Small, K.A. and H.S. Rosen. (1981). "Applied Welfare Economics with Discrete Choice Models," *Econometrica.* 49(1).: pp. 105-130.

Train, Kenneth (1985). "Discount Rates in Consumers' Energy-Related Decisions: A Review of the Literature," *Energy* 10(12): pp. 1243-1253.

Williams, H.C.W.L. (1977). "On the Formation of Travel Demand Models and Economic Evaluation Measures of User Benefit," *Environment and Planning* A,9. pp. 285-344.

Status of Low-Pollution, Energy-Conserving Vehicles in Japan

Dr. Kazuo Kontani

Mechanical Engineering Laboratory
Agency of Industrial Science and Technology
Ministry of International Trade and Industry, Japan

1. Introduction

Supported by the progress of motorization in the post-war era of high economic growth (1960--) and by the expansion of auto exports (1970--), the automobile industry in Japan has continued to grow to its present position.

During this same period, however, environmental problems calling for regulations on exhaust gases, and other difficulties were faced, such as the eruption of two oil crises in succession, which stimulated further efforts to reduce environmental pollution and conserve energy resources. In particular, the regulations on exhaust gases put into effect in 1978 were the most rigid in the world, and still are today. In response to the oil crisis, an energy conservation law was enforced from 1979, and the fuel consumption of vehicles in Japan are now in the top class throughout the world.

As a result of these efforts, the Japanese automobile industry today accounts for 30 % of worldwide automobile production. It has also become one of the major industries inside Japan, responsible for more than 10 % of the total production by all the nation's manufacturers and about 30 % of the entire machinery industry's output.

2. Pollution Reduction Measures

Legal regulations on exhaust gases were first introduced in Japan in 1966. Thereafter, with increases in the number of automobile units and advances in relevant technology, these regulations were progressively tightened. In FY 1978, following the example of the Muskie Law in the United States, the regulations still in effect today were promulgated against gasoline engine passenger vehicles. This law called for a reduction of

NO_x emission to 1/10th of the prior level. Even today, this is the strictest regulation worldwide, and Japan has led the world in reaching the levels stipulated by it.

Exhaust regulations against nitrogen oxides emitted from Diesel engines in trucks or buses were first introduced in 1974, and they have gradually been intensified since then. In view of the urgent need to further reduce NO_x emissions, special efforts are now being made to tighten regulations aimed at reducing emission from heavy Diesel trucks by 15 % by the end of 1990, and revising regulations on light-weight trucks to the equivalent level on passenger cars. Moreover, the Environment Agency is now conducting an investigation to determine a proper framework for regulations on particulate emissions.

The trend in reduction of NO_x emission from gasoline engine and Diesel engine vehicles is shown in Figure 1. It reveals that NO_x emission from gasoline engine passenger cars have been brought down to 8 % of the level prevalent in 1973 before the introduction of regulations, while the levels from Diesel trucks and buses have been reduced to about 40 %.

3. Energy Conservation Measures

Measures to conserve fuel consumption, which originated with the onslaught of two oil crises, one in 1971 and another in 1979, have continued to yield solid benefits. In 1979, an energy conservation law was established which stipulated target values, by vehicle weight, for improving the fuel economy of passenger cars shipped by the makers. (See Table 1.)

Advances are being steadily made in the development of technology to meet these target values and improvements in fuel economy have been reached as shown in Figure 2. Fuel consumption levels for Japanese motor vehicles at present are at the world's highest standard and this fact accounts in part for the popularity of these vehicles in overseas markets. Shown in Figure 3 is the trend in the number of gasoline-powered vehicles owned versus the volume of gasoline consumption, which indicates that the gasoline fuel consumption per vehicle over the last ten years (1977 to 1987) has dropped by 15 %. Consequently, Japan ranks low among the leading nations in terms of gasoline consumption even though the number of vehicles in use has been steadily rising.

4. Alternative Energy Measures

Use of various alternative energy resources has been proposed as one factor in an overall strategy to preserve the environment on a global scale and reduce pollution. In Japan, advances are being made, primarily by the Ministry of International Trade and Industry (MITI), with various R&D targets such as the electric car, the methanol car and the hydrogen car.

Over a six-year period from 1971 to 1976, MITI sponsored an R&D program on the electric car, and continuing efforts have been made since then, but the scope of application for the electric car has been limited mainly because battery performance capacity has not yet reached a level adequate for regular automobile use. Currently, there are several hundred on-road electric cars in use and about 140,000 off-road units such as fork lifts and golf carts.

Considerable R&D has also been done on the methanol vehicle, and it carries not so many difficulties from a technological standpoint. Further steps will have to be taken, however, before this vehicle can be made practical or commercialized. In 1989, the Otto-type methanol vehicle has been subjected to controlled fleet tests by Petroleum Energy Center and JARI, sponsored by MITI, on a small scale. Moreover, from 1990 exposure tests using rats are scheduled to investigate the health effect of unburned methanol, formaldehyde and other substances emitted when the vehicle is started. Before the methanol vehicle can be made practical in Japan, problems in obtaining adequate fuel supplies must first be resolved.

Research on the hydrogen vehicle has been conducted by the Musashi Institute of Technology and by the Mechanical Engineering Laboratory of MITI. At the Musashi Institute, liquid hydrogen has been used, and at MITI, hydrogen is stored in the form of metal-hydrides. There are various technological problems with the hydrogen vehicle, but the major one is that ample supply of hydrogen as a fuel cannot yet be obtained, so this vehicle will not be made practical before the 21st century.

5. Global Environment Aspect

In recent years, attention has focused on environmental problems on a global scale, but these problems cannot be solved by single measures taken independently by each nation. Since the greenhouse effect brought about by the increase in carbon dioxide results directly from energy consumption, the automobile's contribution to the problem cannot be ignored.

Figure 4 illustrates the volume and proportion of CO_2 emissions from various sectors in Japan. Among the technological themes pertaining to the automobile which can be pursued are energy conservation and use of alternative energy sources, but in the near future, the most effective approach will be to pursue significant reductions in energy consumption.

Technological measures which can be taken to reduce energy consumption by automobiles include improved fuel economy through enhancement of engine efficiency and reduction of vehicle body weight. Another powerful factor lies in government comprehensive measures and incentives, examples of which include measures to restructure road environments and coordinate vehicular traffic with other transportation modes so as to improve vehicle driving conditions and measures to modify the tax structure and to promulgate other inducements so that users will shift to smaller, lighter-weight vehicles.

As for introduction of alternative energy sources, there are still many technological problems to be solved, but the most significant obstacle lies in the acquisition of fuel supplies. We must look to multilateral policies by government and extensive controls for the resolution of these difficulties.

Figure 1. Trend in reduction of NOx emission from automobiles in Japan (year in note shows FY of the regulation)

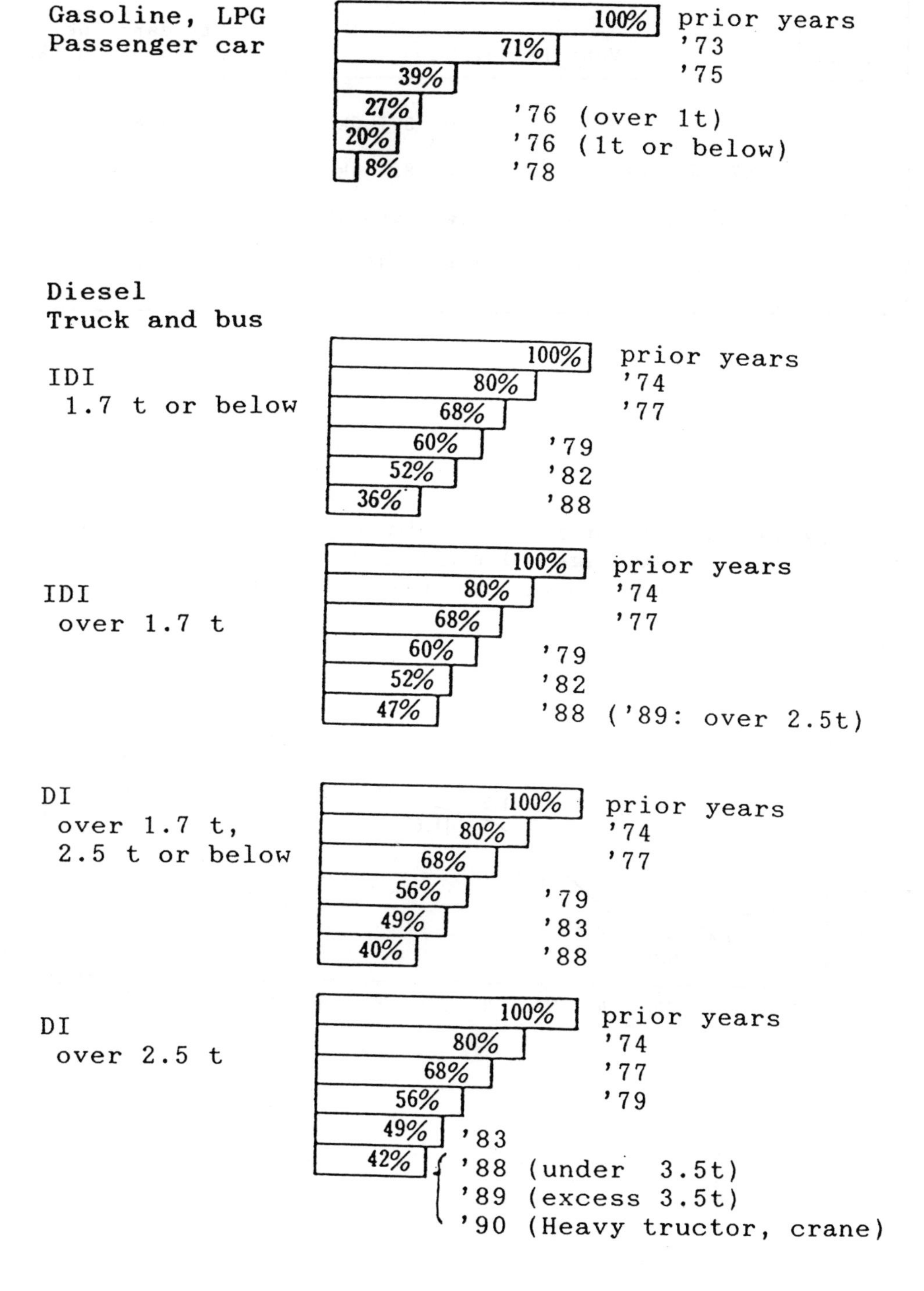

Table 1. Fuel consumption target for passenger cars (energy conservation law of 1979)

Class	Vehicle weight	Standard fuel consumption (km/l)
1	below 577.5kg	19.8
2	577.5kg or above, below 827.5kg	16.0
3	827.5kg or above, below 1,265.5kg	12.5
4	1,265.5kg or above, below 2,015.5kg	8.5

Figure 2. Trend in automobile fuel consumption (Average 10 mode test value for domestic passenger car production)

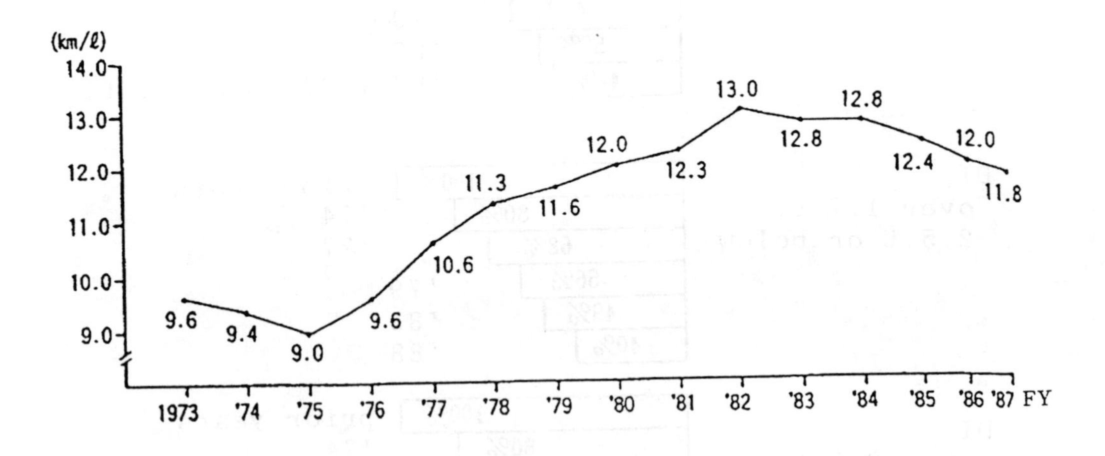

Figure 3.

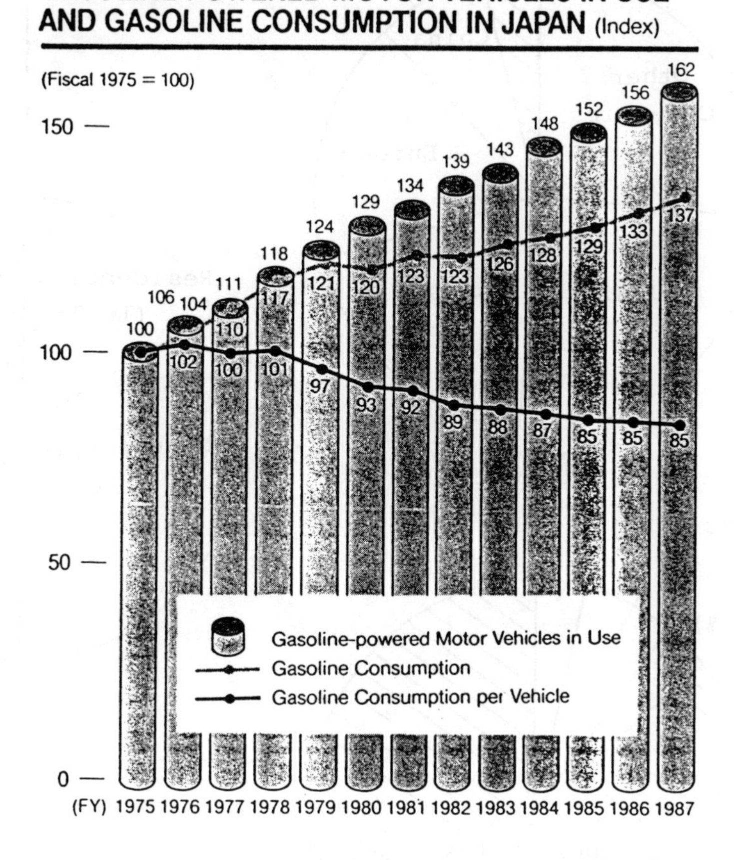
GASOLINE-POWERED MOTOR VEHICLES IN USE AND GASOLINE CONSUMPTION IN JAPAN (Index)
(Fiscal 1975 = 100)
150
100
50
0
100
106
104
111
118
124
129
134
139
143
148
152
156
162
102
110
117
121
120
123
123
126
128
129
133
137
100
101
97
93
92
89
88
87
85
85
85
Gasoline-powered Motor Vehicles in Use
Gasoline Consumption
Gasoline Consumption per Vehicle
(FY) 1975 1976 1977 1978 1979 1980 1981 1982 1983 1984 1985 1986 1987

Figure 4. CONTRIBUTION TO CO_2 EMISSIONS

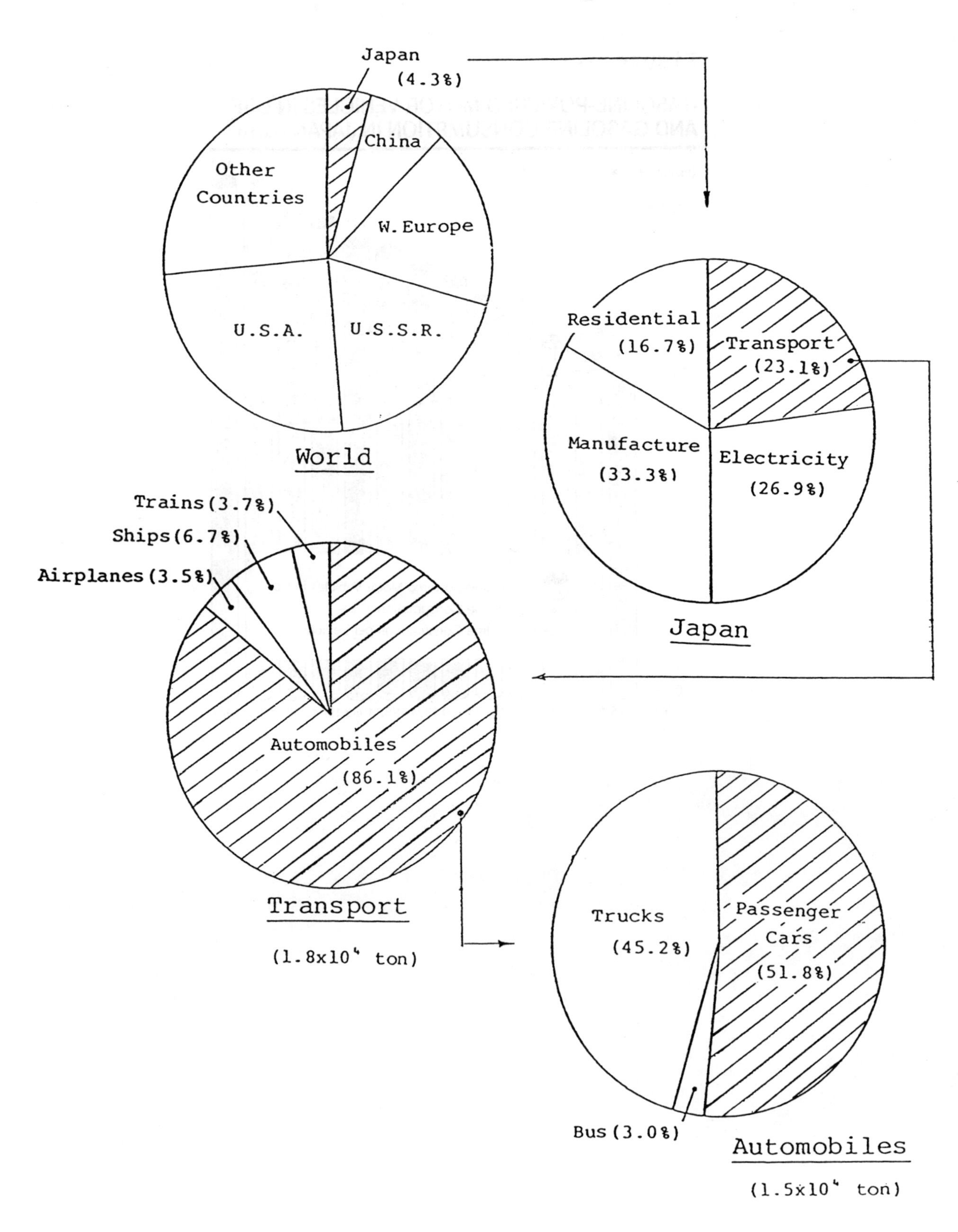

CURRENT STATE OF ADVANCED AUTOMOTIVE TECHNOLOGY IN UK

J F Bingham
IC Engines Division
National Engineering Laboratory
East Kilbride
Glasgow

1 INTRODUCTION

This review surveys briefly the current state of technology in the UK as it affects fuel economy and emissions of automobiles. Firstly some national trends are identified and then current technological developments are summarised.

2 TRENDS

The dominant factor in the vehicle market is the need to produce a car acceptable to the consumer. Influences on the market in the UK include:

- Consumer's expectation of good performance.
- High level of vehicle attributes desired.
- Widespread provision of company cars sustains the market for executive and luxury cars.

Although there is a growing awareness of environmental issues the catalytic convertor option is still not widely taken up. It is also significant that sales of unleaded gasoline remained low until the government introduced a more favourable price structure.

Within the industry there is a conviction that legislation should not inhibit technology. There is some disappointment that the continued development of lean burn engines has been arrested by tighter emissions standards, since combustion of lean air/fuel mixtures would have helped to reduce CO_2 levels.

The general trends in fuel economy improvement over the past decade are shown in Figures 1 and 2.

3 TECHNOLOGICAL DEVELOPMENTS

In the period up to the mid 1970's the adaptation of engines to meet US emissions standards resulted in a severe penalty in power output. As emissions legislation has become more universal, technology has been developed to recover this lost performance while meeting pollution limits. Specific power output of engines has increased.

Some companies are moving towards a total energy policy in which the energy implications of manufacturing strategy and production methods are considered as well as energy consumption by vehicles.

These general technological trends have had specific effects on:

i Powertrain;

ii Vehicle body.

3.1 Powertrain

Trends in spark ignition engine design include:

- Multi-valve cylinder head design.
- Use of MPI and TBI fuel systems.
- Development of variable geometry intake systems.
- Design of high turbulence combustion chamber (necessary for lean burn combustion).
- Use of EGR to dilute charge and enable engine to run on lean gas/fuel ratios while maintaining an air/fuel ratio of stoichiometric.
- Development of high energy ignition systems.
- Design changes to exhaust system to reduce time to catalyst light-off.
- Improved control strategies for transient fuelling of engines including adaptive and self-optimising systems.
- Use of in-cylinder sensor to monitor engine operation.
- Renewed interest in 2-stroke engines.

Diesel engine developments include:

- High speed di design.
- Electronic control.
- 2 stage injectors.

Problem areas remain such as effective particulate traps and NO_X levels.

Evolution of advanced transmission systems include:

- More ratios - manual and automatic.
- CVT.
- Four wheel drive systems.

3.2 **Vehicle Body**

Developments in this area include:

- Reduced weight. This conflicts with refinement and safety requirements so a compromise is required.
- Improved aerodynamics.

4 FUELS AND LUBRICANTS

In response to pressures of energy conservation and pollution, research is in progress on various alternative fuels including methanol. The effect of lubricants on emissions is also a topic of interest.

5 UK ENGINE EMISSIONS CONSORTIUM

The United Kingdom Engine Emissions Consortium (UKEEC) is a joint venture between the UK Government and Industry, with the aim of improving vehicle exhaust emission technology.

Increasingly severe environmental regulations on vehicle exhaust emissions in Europe and North America were a major factor in launching the programme.

The UKEEC has been established on the advice of the Department of Trade and Industry's former Vehicle Advisory Committee (VAC) which recognised the benefits which would flow from the pooling of knowledge and resources, and recommended that the DTI should provide encouragement and financial support.

The Consortium has drawn together some twelve industrial companies who include diesel and petrol engine manufacturers, component suppliers and firms in the oil and chemical industries. Research organisations and UK universities are also involved as subcontractors. The Department of Trade and Industry is providing some £2.3 million funding over a three-year period.

The Consortium is currently working on eleven projects. Five of these are related to petrol engines and six to diesels. Each is supported by at least three collaborating companies. The longest project is expected to last three years, the shortest one year.

The projects themselves cover a wide variety of subjects such as:

- New concepts in engine and component design.
- Electronic controls and sensors.
- New techniques in engine development.

- The effects of new oil and fuel forumulations.

- The use of non-fossil fuel.

The projects are being carried out in addition to work being done by the firms concerned as part of their individual research and development programme.

The projects are at the leading edge of emissions technologies and are certain to set the directions for future research and development in the industry. They will also provide an important platform for further collaboration with our European partners.

The members of the UKEEC are:

Perkins Engines Group Ltd (the manager company)
Exxon Chemical Ltd
Ford Motor Co Ltd
Garrett Automotive
ICI Chemicals and Polymers
Jaguar Cars Ltd
Johnson Matthey plc
Lotus Cars Ltd
Lucas Automotive Ltd
National Engineering Laboratory
Rover Group Ltd
T & N Technology Ltd

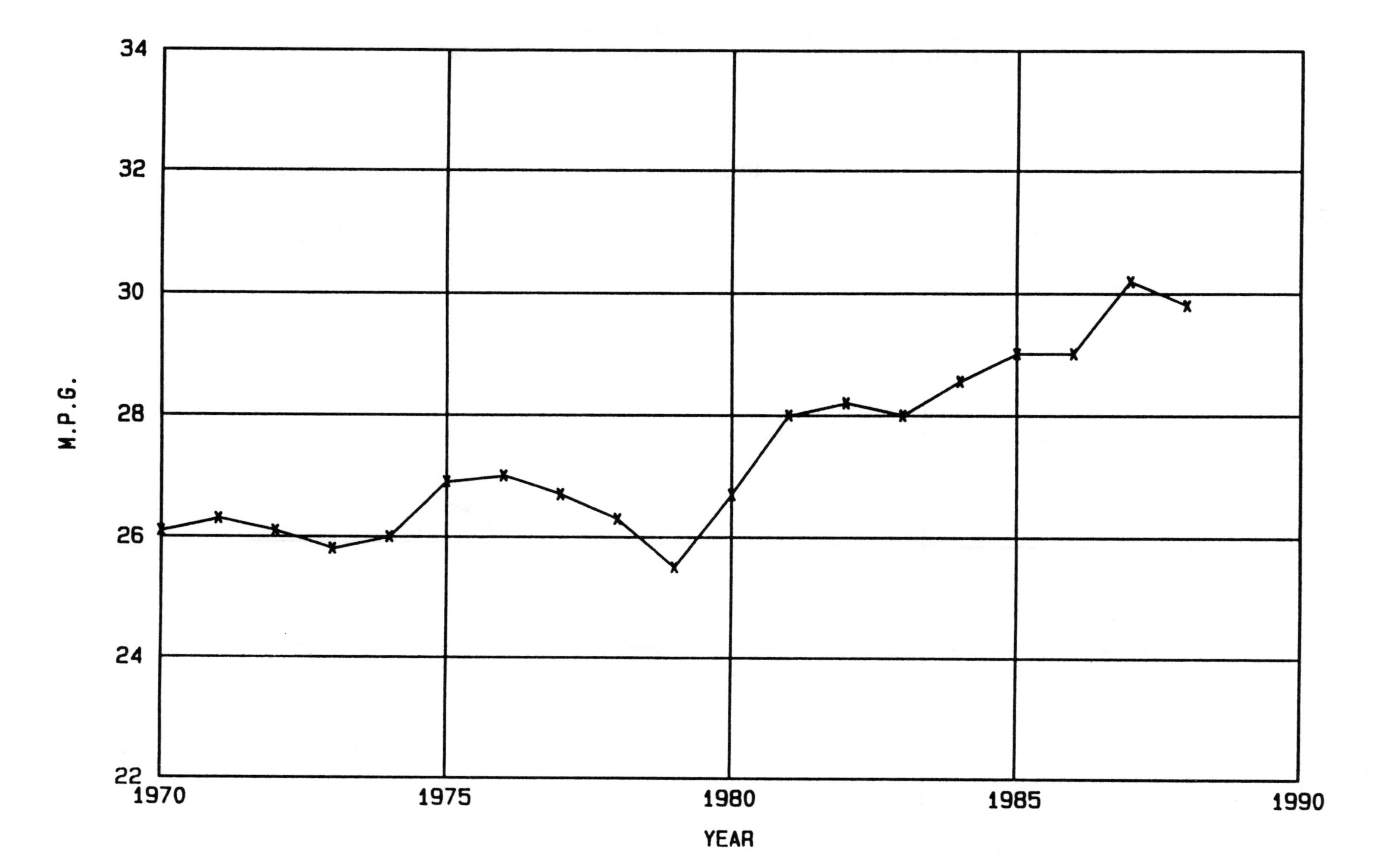

FIG. 1 FUEL CONSUMPTION IN UK - ALL GASOLINE ENGINED VEHICLES

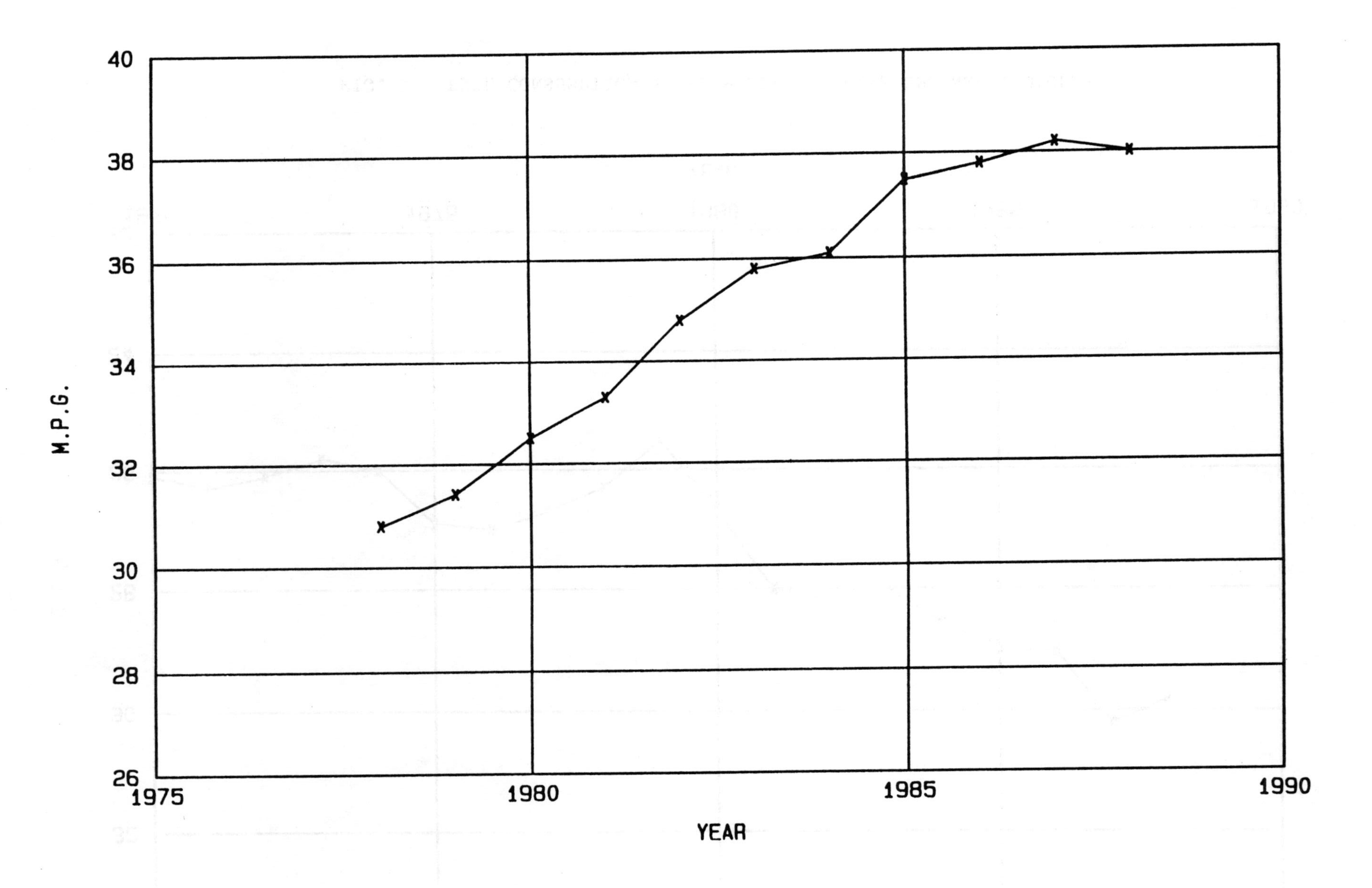

FIG. 2 OFFICIAL NEW CAR MILE/GAL FIGURES FOR UK (REGISTRATION AND ROAD TYPE WEIGHTED)

STATUS AND POLICY OF ENERGY SAVING

WITH RESPECT TO MOTOR VEHICLES IN JAPAN

Takuro MIYAZAKI

Office of Motor Vehicle Pollution Control
Engineering and Planning Division Land Transport Engineering Department
Ministry of Transport Government of Japan

1. Experimental Calculation on the Amount of Carbon Dioxide Emission from Motor Vehicle Transport in Japan

As for energy saving of motor vehicles, various measures have been taken in every occasion of oil shortage crisis. In the case of carbon dioxide, a major element of greenhouse gases which calls growing concern in the world, the measures to reduce the emission are quite limited and the knowledge to reduce carbon dioxide emission is very poor up to now. Therefore, the measures to reduce carbon dioxide emission are very important in considering further measures for energy saving of motor vehicles. Table 1 is an example of calculated results of carbon dioxide emission from motor vehicle transport.

(1) The total amount of carbon dioxide emission in Japan shares 4.3% of the emission in the world. The percentage of tranport sector is 23%, approximately 1/4 of the total emission in Japan.

(2) The amount of emission from motor vehicles covers 86% of total amount of transport sector, which indicates that considerable amount is due to motor vehicle transport. In the tranport sector, the share of 44% is for passenger cars, 3% for buses, and 39% for trucks.

(3) The total amount of carbon dioxide emission from motor vehicles, which is deemed 100% for 152 million tons, is divided by 52% of passenger cars for 79 million tons, 3% of buses for 5 million tons, and 45% of trucks for 69 million tons, individually.

2. Present Situation and Policy of Energy Saving of Motor Vehicles

2.1 Energy Saving Act

In order to utilize effective use of petroleum resources, a law with a title of "Act for the Rationalization of Use of Energy", so called "Energy Saving Act", which contains certain measures to rationalize the use of energy in the area of factories, buildings and machinery such as motor vehicles, was proclaimed on 22nd June and enforced on 1st October in 1979. It is provided in the Act that the government shall specify the judgement standard of motor vehicle fuel efficiency, and that the information on the fuel efficiency shall be supplied to motor vehicle users.

According to the Act, motor vehicle manufacturers and importers have to develop motor vehicles with better fuel economy, and to indicate fuel efficiency. The government shall demand to report necessary matters from motor vehicle manufacturers and importers and shall investigate them with respect to the improvement of fuel efficiency. Besides, the government shall recommend to improve fuel efficiency and the way of indication of fuel efficiency.

2.2 Judgement Standard of Fuel Efficiency of Motor Vehicles

As to the judgement standard of motor vehicle fuel efficiency, a miniterial ordinance entitled "Ordinance for the Calculation of Motor Vehicle Fuel Efficiency" was announced on 27th December 1979. At the same time, an announcement was published with a title of "Judgement Standard of Motor Vehicle Manufacturers and so forth to Improve Motor Vehicle Performance". The announcement specified the targets for type-designated gasoline-fueled passenger cars that each motor vehicle manufacturer and importer should accomplish in their production or import according to their vehicle weight. (See Table 2.)

Although the standard was accomplished in the target year by each manufacturers, the government is still requesting the manufacturers to report the status of improvement of fuel efficiency after the target year, and it has concern about the situation of improvement. The fuel efficiency of each model of passenger car is still improving, but the total fuel efficiency of each motor vehicle manufacturer has shown somewhat decline recently, because of heavier weight and bigger size of recent passeger car models as well as the introduction of power-assisted systems such as automatic transmission system and power-assisted steering system.

2.3 Energy Saving Measures of Motor Vehicles

In accordance with the Act, following measures are taken to promote the energy saving of motor vehicles.

(1) Administrative guidance to motor vehicle manufacturers: Necessary guidance is made to motor vehicle manufacturers for the improvement of fuel economy.

(2) Popularizatin of motor vehicles with better fuel economy: Necessary guidance is made to the motor vehicle manufacturers and importers so that they will indicate, in the vehicle catalogue and on the panel in the showroom, the fuel efficiency of type-designated gasoline-fueled passenger cars as well as their vehicle name, model, engine displacement and the kind of power transmission system. Moreover, from 1976, the government has come to announce publicly the fuel efficiency of motor vehicles, at the time of type designation, for every type-designated gasoline-fueled passenger cars. The government has also editted and published "An Outlook of Passenger Car Fuel Economy" in order to popularize the vehicles with better fuel economy.

(3) Investigation of vehicle manufacturers: In order to grasp the situation of the matters listed in item (1) and (2), the government has the authority to get report on the relevant matters, and to make investigation of factories and so forth if necessary. With the result, the government is able to recommend to manufacturers with regard to the vehicle performance and the indication of fuel economy.

(4) Education of vehicle users: In order to spread widely the driving technique and inspection/maintenance which are useful for improvement of fuel economy, the

government has requested to the cooperative organizations to make public relation activities to vehicle users with booklets. In addition to the request, the government has requested to relevant authorities to notify the issue to vehicle operators and vehicle maintenance authorities in the case of mandatory trainning.

3. Research and Development for the Promotion of Energy Saving with Respect to Motor Vehicles

3.1 Urgent Items of Research and Development for Energy Saving

The object of the Energy Saving Act is to endeavour measures for rationalization of energy use taking into consideration of the energy situation in Japan. However, energy saving issues nowadays is becoming important from the new standpoint of global environment issues such as greenhouse gases like carbon dioxide. Therefore, future policies must include the evaluation and estimation of improvement measures for motor vehicle energy efficiency as well as the measurement technique of the efficiency. Based on those results, the element techonologies for motor vehicle energy improvement should be promoted.

3.1.1 Developmennt of Measuring Technology of Motor Vehicle Fuel Consumption

The measurement and evaluation techniques, which are important for the development and evaluation of fuel-economy vehicles, have to be developed. Some examples of items for consideration are shown in the following.

(1) development of high accuracy measuring technique of fuel consumption
(2) development of measuring mode and test method on chasis-dynamometer which are closely related with the actual usage of vehicles

3.1.2 Developmennt of Element Technology for Energy Saving of Motor Vehicles

The indivisual elements of technology which are used in engines and bodies for energy improvement must be researched and developed. Besides, the evalution method, technique to maintain the performance, and its collective technology of the element technologies must be researched. Some examples of items for consideration are shown in the following.

(1) control technology of lean-burn system
(2) moderate control technology with the application of electronics
(3) evalutaion of the influence of light body and structual strength
(4) measures to reduce running losses
(5) development of ultra-micro vehicles
(6) development of techniques to find failure and function deterioration for in-use vehicles applying artificial intelligence

3.2 Research/Development and Evaluation of Performance with Respect to New System Vehicles with Alternative Fuels

In order to promote energy saving and to reduce substantial amount of carbon dioxide emission, a positive turn to altenative fuel have to be considered in addition to the existing measures of energy-saving. The utilization of the vehicles with new systems, such as methanol-fueled, hybrid-powered and electric-powered, must be promoted. Furthermore, suitable method evaluating the utility, fuel economy and low-pollution chracteristics must be established. Some examples of items for consideration are shown in the following.

(1) energy regeneration technology
(2) control technology of engine with alternative fuels
(3) pollution analysis of alternative engines (unregulated substances, application technique of catalysts)
(4) preparation of supply system of alternative fuels
(5) financial assistance to promote popularization (support of development budget, preferential treatment in tax system)

4. Future Prospect of Energy Saving Measures for Motor Vehicles

The future prospect of energy saving measures for motor vehicles at present, with respect to the improvement techonology of fuel efficiency using gasoline or light oil which is used as petroleum fuel and the future technology of energy saving including the use of alternative fuels, is shown in Table 3 as a personal viewpoint.

Table 1 Result of Experimental Calculation on the Carbon Dioxide Emission from Motor Vehicle Transport in Japan

Item		Percentage	Amount of emission (million tons)	Year/Origin
total amount of emission in Japan		* 4.3%	* 230	1985/White Paper from Environment Agency
total	public welfare	16.7%	130	1986/Environment Agency
	production	33.3%	160	
	electric power	26.9%	210	
	transport	23.1%	180	
	total	100.0%	780	
transport	railway	3.7%	6.5	1986/Energy Policy Division, Ministry of Transport
	marine transport	6.7%	11.8	
	air traffic	3.5%	6.2	
	motor vehicle	86.1%	151.9	
	(passenger car)	(44.6%)	(78.7)	
	(bus)	(2.6%)	(4.6)	
	(truck)	(38.9%)	(68.6)	
	total	100.0%	176.4	
motor vehicle	passenger car	51.8%	78.6	
	bus	3.0%	4.6	
	truck	45.2%	68.7	
	(private use)	(27.8%)	(42.3)	
	(public use)	(17.4%)	(26.4)	
	total	100.0%	151.9	

Note: 1. Calculated by the Energy Policy Division, Ministry of Transport in December 1989.
2. * marks indicate that the weight is expressed in carbon equivalent, while others in carbon dioxide.

Table 2 Judgment Standards for Motor Vehicle Fuel Efficiency

vehicle weight (kg)	less than 577.5	577.5 or more and less than 827.5	827.5 or more and less than 1,265.5	1,265.5 or more and less than 2,015.5
target value by 10-mode (km/ℓ)	19.8	16.0	12.5	8.5
		13.0 (in case that both categories are produced)		

Table 3 Future Prospect of Energy Saving Measures for Motor Vehicles

Item	Measures	Effectiveness	Expected period of application
1. Improvement of fuel efficiency	(1) Setting-up of new target value for fuel efficiency (use of Energy Saving Act)	very effective	short-term
	(2) Introduction of gasoline-fueled lean-burn engine with full electronic control	effective	short-term
	(3) Development and popularization of hybrid power unit (city bus with diesel and inverter-alternator unit)	effective	short-term
	(4) Development and popularization of regenerating unit (retarder-regenerator brake)	effective	short-term
	(5) Development of ultra-mini-size vehicle (two-wheeled or three-wheeled vehicle)	effective	medium-term
	(6) Development and popularization of variable element energy-saving type engine with electronic control (air-fuel ratio, variable use of cylinders, valuable compression ratio)	effective	medium-term
	(7) Development of high efficient engine (ceramic engine)	effective	long-term
2. Change to alternative fuel	(1) Popularization of vehicles with liquefied petroleum gas engine (already popular in taxis)	none (effective for the reduction of CO_2)	short-term
	(2) Popularization of methanol-fueled vehicles	none (effective for clean air)	short-term
	(3) Development and popularization of vehicles with compressed natural gas	less effective (effective for the reduction of CO_2)	medium-term
	(4) Development and popularization of electric vehicles	less effective	medium-term
	(5) Development of hydrogen-fueled vehicles	effective	long-term

Note: 1. short-term; 3-5 years
2. medium-term; 5-10 years
3. long-term; more than 10 years

ENERGY AND EMISSIONS ADAPTATION OF A ROAD TRANSPORTATION SYSTEM - THE SWEDISH EXAMPLE

Discussion Note prepared by:
Gunnar Kinbom, Swedish National Board for Technical Development
Ragnar Thörnblom, Swedish Transport Research Board

It is well understood in Sweden as well as in many other countries that, due to its emissions, the transportation sector has a major local, regional and global impact on air in the environment. As a consequence of the growth of vehicle density and traffic volume, and simultaneously the reduction of harmful emissions from other sources, such as industrial processes, domestic space heating systems etc., the transportation sector's share of total emissions is becoming increasingly higher. To achieve a long-term reduction of the total emissions, actions in the transportation sector are becoming increasingly important.

The present situation (January 1990) in Sweden can be described as follows:

- The number of vehicles (vehicle density) increases
- The average driving distance per vehicle increases
- The average fuel consumption tends to increase
- Traffic james in urban areas create increasing problems
- The pace of traffic is becoming increasingly harder and riskful
- The price of fuel, in real terms, is lower than ever.

The traffic environment debate and decisions concern the following issues:

- The catalytic converter is generally mandatory in gasoline-driven vehicles
- The share of the emissions from diesel vehicles is growing
- The Parliament's target for reducing nitrogen oxides and carbon dioxide emissions will not be achieved given present traffic development
- A system for environmental taxes on nitrogen oxide, sulphur dioxide and carbon dioxide is on the way to being introduced
- The so-called greenhouse effect has increasingly come into focus in the energy and environmental debate.

A number of Swedish agencies assigned by the Government have performed a collaborative study on an energy and environment adapted transportation system. The study, to be completed in March 1990, will constitute the basis (background material) for a Parliament bill on transportation and the environment.

So far, we can state that:

- Given the present development, the target for reducing nitrogen oxide emissions to 30% by the year 1995 will not be achieved. A maximum of a 50% reduction can be achieved at a later date. Carbon dioxide emissions will have increased by 20% in the year 2015, in spite of the target, according to which no increase is allowed. If economic growth proves to be higher than current forecasts, the emissions from the traffic sector will rise sharply.

- Emission reductions can be achieved by enacting more stringent requirements on vehicles, and through various measures directed to curb the demand for transportation. The enaction of the so-called California requirements would mean that, although the nitrogen oxide target would be achieved, the carbon dioxide target would not.

- The target for reducing carbon dioxide can be achieved if stringent regulations for vehicles and measures to curb transportation demand are combined with sharply higher fuel prices and an expanded public transportation system.

- Agricultural tractors and machines, shipping and air traffic must be subject to stringent emission limits in the same way as road traffic, if the target levels are to be accomplished.

- A serious discussion is under way in Sweden about tightening the emission target for nitrogen oxides and carbon dioxide even further than today. The consequences of such an action are that the transportation system must be altered and partially reoriented towards building new infrastructures in the transportation system, and incorporating comprehensive technological changes in the vehicles.

A special study on vehicle technology has been made in which the technical alternatives and the potentials for these are examined. In this study, the goals for various levels of carbon dioxide emissions determine the selection of the technology. The outlook is 25 years into the future, that is the year 2015.

Based on the study, the conclusions are:

- Seen in a regional and local perspective, there presently exist technical solutions, the implementation of which will require moderate efforts even for very stringent requirements for low levels of harmful emissions.

- Depending on the emission level target in the global perspective (primarily carbon dioxide emissions), energy from renewable energy sources must be utilised to a more or less large extent. This requires the application of alcohol, or electric operation, etc.

Development and demonstration programmes for vehicles using improved or new technology are under way in Sweden. Examples of the technologies applied in these programmes are:

- New fuels and exhaust gas cleaning technologies for (diesel) operation on diesel oil;

- Operation of both heavy and light vehicles on methanol and ethanol;

- Operating of heavy vehicles on LPG, natural gas, biogas and hydrogen;

- Operation of both heavy and light electric vehicles on batteries or hybrid systems with a combustion engine charger on board.

The implementation of changes in a traffic system with very ambitious energy and environmental goals will require that all of the actors in the market interact and share a common approach to the problems.

Collaboration is necessary on the international level which, in its initial stage, will lay the groundwork for a common approach to the problems and will subsequently be the basis for a joint strategy in order to achieve mutual goals. The common approach and enforcement of the same requirements over national boundaries is necessary to be able to create markets for new technologies in the traffic system.

Assessment of Potential Gains in Fuel Economy:

the U.S. Experience

Donald Bischoff
Director, Office of Regulatory Analysis
National Highway Traffic Safety Administration
U.S. Department of Transportation

A. ASSESSMENT OF FUTURE POTENTIAL GAINS IN FUEL ECONOMY FROM TECHNOLOGY

Recent debate in the United States about further gains in fuel efficiency from technology have centered on a report prepared for the Department of Energy (DOE) by Energy and Environmental Associates (EEA). EEA has developed projections of fuel economy for the years 1995 and 2001 for three separate scenarios, representing different technologies and markets:

- Scenario 1 is the "product plan," which assumes no new legislation.
- Scenario 2 is a "maximum available technology" scenario which is probably the maximum possible fuel economy level attainable if manufacturers are forced to do so, starting now.
- Scenario 3 is a hypothetical maximum case similar to Scenario 2, but also holding product utility constant at 1987 levels.

The EEA report estimates a possible 17.1 percent CAFE improvement between 1987 and 1995 for Scenario 1. For the period 1995 to 2001, EEA estimates additional improvements of 9.9%, 17.2% and 27.6% respectively for Scenarios 1, 2, and 3. The U.S. auto manufacturers claim that the EEA study grossly overstates gains in fuel economy associated with various technologies. Where EEA estimates a 17.1 percent improvement in the near term, the companies (Ford, G.M. and Chrysler) estimates are between 7.1 percent and 8.6 percent. The companies' estimates are claimed to be based on "engineering analysis." For the 1995 to 2000 time frame, the companies' estimates range from 2.8 to 3.4 percent improvement for Scenario 2 and 13.8 to 14.6 percent improvement for Scenario 3.

The EEA report estimates light duty truck fleet fuel economy gains from 1987 to 1995 as 9.5 percent and 12.6 percent for Scenario 1 and Scenario 3 respectively. For 1995 to 2001 the increases are estimated as 11.1 and 16.2 percent respectively for Scenarios 1 and 3. The domestic auto industry has not publicly commented on the light duty truck estimates.

The NHTSA has also supplied comments to DOE on the EEA report. A major concern with the report is the considerable increase in fuel economy levels for the 1995 product plan scenario, despite a lack of current market place demand for fuel efficient vehicles. The agency also believes that the report overstates the gains from some of the near term technologies.

B. SHOULD CAFE STANDARDS BE RAISED?

Congress passed the Energy Policy and Conservation Act (EPCA) in the wake of the 1973-74 OPEC oil embargo crisis as a major step toward energy independence. The Act set up a schedule of passenger car fuel economy standards to be achieved by automobile manufacturers, culminating at 27.5 miles per gallon (mpg) for model year 1985 and thereafter. The Secretary of Transportation is empowered (but not required) to amend that standard for any model year to the "maximum feasible" fuel economy attainable by the industry for that year, taking into consideration:

- economic practicability
- technological feasibility
- the effect of other motor vehicle standards
- the need of the nation to conserve energy.

A manufacturer that fails to meet the standard is liable for a civil penalty equal to $5 for each one tenth mpg its CAFE is below the standard, multiplied by the total number of cars the manufacturer sold in the model year. A manufacturer can earn credits at the same rate for exceeding the standard. Credits can be carried forward or backward for up to three years to offset nonattainment in other years.

Congress was concerned that a manufacturer might elect to fill out the small car end of its product line with captive imports to meet the CAFE standard rather than develop more fuel efficient domestic models. To avoid this potential loss of jobs and ensure that the industry developed domestic models with high fuel economy, Congress required manufacturers to meet CAFE standards separately for their domestic and imported fleets. In fact, it did not develop that way. Although the U.S. manufacturers all offered domestic subcompact models, their concentration has been on the higher profit margin larger vehicles, with increasing reliance on captive imports to fill out the lower end of the product line.

Because of a combination of the Act and market conditions, fuel economy of new cars improved from a low of 14.2 mpg in 1974 to 28.5 mpg in 1987. The domestic new car average fuel economy rose from 12.8 mpg in MY 1974 to 25.0 in MY 1982 as the manufacturers exceeded the CAFE standards and accrued credits. Thereafter, with a strong domestic economy, falling fuel prices, and a dependable fuel supply buyer preference shifted toward heavier cars with higher performance and GM and Ford were unable to keep up with the increasing standard.

DOT, in evaluating the changed market conditions, found that the manufacturers had made a reasonable effort to comply and, in response to petitions, lowered the standards for the model years 1986 through 1989. In 1988, Ford estimated a CAFE of 26.4 mpg and GM got up to 27.6 mpg, but projected an ability to achieve only 27.2 mpg in 1989. For MY 1990, the DOT has retained the statutory standard of 27.5 mpg, based on assessment of the manufacturers' plans and an evaluation of the statutory criteria, especially the need for the nation to conserve energy.

While the CAFE statute may have had some impact in encouraging improved fuel economy (along with other market forces), the current statutory scheme has significant deficiencies. First, the CAFE statute addresses only the supply of fuel-efficient vehicles. There is little to promote increased demand for those vehicles. Although currently pending legislation being considered in the U.S. Congress reflects concern about energy security and global climate change, consumers have been confronted with the countervailing effects of low fuel prices. Second, there is the relative benefit enjoyed by most Asian manufacturers under the law by virtue of their concentration on smaller vehicles. The CAFE standards have typically been at or slightly above the CAFES of the domestic manufacturers, but well below those of the Asian manufacturers. Third, there is a perverse incentive for the domestic manufacturers to "improve" the CAFES of their domestic fleets by shifting the production of their larger, less fuel-efficient cars abroad, so that those cars can be averaged together with the small cars in their import fleets.

In addition, since a manufacturer's CAFE is largely dependent on the mix of vehicles it sells, a manufacturer with a compliance plan whose success hinges on market assumptions (e.g., consumer demand for small vehicles, the price of gasoline) that prove incorrect, may be able to comply with CAFE standards only by restricting the sales of its less fuel-efficient vehicles.

Recent studies (OTA, EPA, DOE) indicate that it maybe technologically feasible to achieve higher levels of fuel economy in the mid-nineties-30 mpg with current trends and 40 mpg with government intervention. DOT does not necessarily agree, since no agency has yet analyzed the economic practicability for full-line manufacturers to produce and sell a mix of cars that would achieve markedly higher corporate average fuel economy. If the Administration decides that substantial improvements in fuel efficiency are needed as part of an effort to ensure long-term energy security and/or to address global climate change, and that exclusive reliance on the market would not secure those improvements, decisions must be made as to the best means of achieving those improvements consistent with preserving the auto industry.

The question of how to make further gains in fuel efficiency will require a consideration of a multitude of factors such as economic practicability and technological feasibility as well as the need for energy conservation. If it is determined that higher CAFE standards are necessary and desirable it is imperative that the flaws in the CAFE mechanism be corrected. Otherwise, it is likely that the increased CAFE standards will seriously impair the competitiveness of full-line manufacturers and will not likely produce the expected energy savings.

C. EFFECTS OF A FLEET OF LIGHTER CARS

Comparisons of fatalities per registered vehicle show that lighter cars have more fatalities per car than do heavier cars. The differences are greatest for multiple vehicle accidents and for single vehicle rollover accidents.

Another conclusion from ten years of fatality data is that fatalities per registered vehicle have declined within each car weight class. This suggests that occupants of cars of all weight classes are benefiting from other improvements and that these improvements are large enough to compensate for an increase in injuries from the shift to lighter cars.

A difficulty with using this parameter (fatalities per vehicle) is that it depends on numerous factors, including crashworthiness, vehicle propensity to crash involvement, and differences in vehicle use and drivers. To date, NHTSA has examined the vehicle crashworthiness aspect in two types of accidents: single vehicle non-rollover and multiple vehicle. The agency has also investigated vehicle propensity for crash involvement in rollover accidents.

Detailed accident data show that the number of driver moderate injuries, per vehicle which had to be towed from the scene, declines with increasing car weight, after controlling for damage type, number of involved vehicles, rollover occurrence, severity of non-rollover crashes and victim age. The moderate injury rate in single vehicle accidents is 1.5 times as high in mini-compacts as in the largest cars; it is 1.33 times as high in multiple vehicle accidents. Serious injury rates (AIS $\geq$ 3) are 1.22 times as high in mini-compacts as in the largest cars in single vehicle accidents and 1.17 times as high in multiple vehicle accidents after adjusting for differences in crash type, crash severity and victim age.

Moderate injury rates (AIS = 2) decline with increasing car weight in both single vehicle and multiple vehicle nonrollover accidents. This is the case for nonrollovers as a group, and specifically for frontal, left, and right side impacts. Fatality rate is not a valid measure of crashworthiness for rollovers. Cars that tend to rollover easily (small, narrow cars) do so in crashes of intrinsically low severity. These rollovers have low injury rates.

Between 1980 and 1987, the average weight of the on-road fleet dropped from 3524 to 3138 pounds. Models of driver moderate injury rates as a function of car weight produce estimates that this 386 pound weight drop led to an increase of 5.6 percent in the number of moderate injuries in single vehicle non-rollover collisions. The agency is currently conducting similar analyses on two car collisions. Preliminary indications are that all crash modes (head-on, side and rear-end) show a

tendency toward higher injury rates in crashes with lighter cars, although the statistical relationships are weak. These analyses do not suggest the mechanism for this persistent relationship, and there is considerable variability in the data.

Agency analyses show that narrower, lighter, shorter cars have higher rollover rates than wide, heavy, long ones under the same crash conditions. During 1970-82, as the market shifted from large domestic cars to downsized, subcompact or imported cars, the fleet became more rollover prone. That may have been partly offset by increases in the track width of some imported cars after 1977. The net effect of all car size changes since 1970 is an increase of approximately 1340 rollover fatalities per year. This includes 1220 lives lost per year due to a shift to subcompact and imported cars, 350 lives lost per year due to downsizing of existing car lines from 1975-82 and 230 lives saved due to wider tracks for some imported cars 1977-82.

D. PENDING LEGISLATION IN THE UNITED STATES

A number of CAFE-related bills have been introduced in the U.S. Congress. The bills follow one of three general approaches:

1. Increasing each manufacturer's CAFE level by a percentage of its model year 1988 or 1989 CAFE,

2. Increasing the CAFE standard to a specified level for all manufacturers, and

3. Requiring stringent emissions standards for carbon dioxide that can be met only by achieving higher CAFE levels.

Several bills utilize the "percentage increase" approach. Some would require percentage increases as high as 45 percent by 1995 and 65 percent by 2001. The bill furthest along is Senator Bryan's (S. 1224), which mandates somewhat lower increases. Under the Bryan bill, passenger car manufacturers must raise their CAFEs 20 percent (using a MY 1988 base) by 1995, with a maximum standard of 40 mpg, and 40 percent by 2001, with a maximum standard of 45 mpg. Light truck manufacturers would have to raise their CAFEs by the same percentages, with caps of 30 mpg for 1995 and 35 mpg for 2001.

A bill introduced by Senator Metzenbaum would raise the standard in stages to 34 mpg by 1996, with a provision to prohibit backsliding by any manufacturer that already exceeds the level of the standard. This bill has been referred to Senator Bryan's subcommittee, where it is likely to be superceded by Bryan's bill.

The Clean Air Act amendments reported out of the Senate Committee on Environment and Public Works (S. 1630) contain limits for carbon dioxide tailpipe emissions that would require each manufacturer to have an average fuel economy of approximately 34 mpg for 1996 and 41 mpg for 2000.*

In early March 1990, Senate and White House negotiators agreed to remove the carbon dioxide standards from the Senate version of the Clean Air Act amendments, in an attempt to develop a compromise bill that would be signed by the President. In addition, attempts by Senator Bryan to attach his proposal to the Clean Air Act amendments have been successfully opposed by supporters of the compromise amendments. However, the Senate Majority Leader has stated his intention to bring a comprehensive global warming bill to the Senate floor this year. The CAFE proposals will likely be considered as part of this bill.

On March 7, 1990, Secretary of Transportation Skinner sent a letter to the Chairman of the Senate Committee on Environment and Public Works stating the Administration's opposition to the Bryan bill. In the letter, the Secretary noted that: (1) the technical feasibility of standards at the levels in the bill, without significant vehicle downsizing, has not been demonstrated; (2) the proposed CAFE increases would radically curtail the choice of new vehicles available to consumers; (3) this would result in lost auto sales and increased unemployment; and (4) there would be a noticeable adverse impact on highway safety. Furthermore, the Secretary noted that the Federal government is placing major demands on the auto industry through additional safety and emissions standards. The Secretary stated that this is not the appropriate time to impose major new CAFE requirements. Opponents of the Bryan proposal read the Secretary's letter into the record during the debate that preceded Bryan's withdrawal of his amendment.

*The bill contains carbon dioxide standards, to be met on a fleet-average basis, of 262 grams per mile for 1996-99 and 220 grams per mile for 2000 and subsequent model years. These were translated into miles-per-gallon figures by using the assumption that the combustion of one gallon of gasoline results in emissions of 19.7 pounds of carbon dioxide.

THE AUSTRALIAN EXPERIENCE IN DEVELOPING A STRATEGY OF LOWER EMISSION AND LOWER FUEL CONSUMPTION IN MOTOR VEHICLES - PROBLEMS, LESSONS AND BENEFITS

John Tysoe
Commonwealth Department of
Primary Industries and Energy
Canberra, Australia

BACKGROUND

In the early 1970s, the upward trend in ozone levels in Sydney was identified as a cause of concern on environmental and health grounds. The National Health and Medical Research Council had recommended as a long term goal that the number of days in a year that the maximum hourly average ozone concentration was above 0.12 ppm (240 mg/m^3) should not exceed one. It was agreed that the most effective means of controlling ozone was to limit its precursor, hydrocarbon emissions to the atmosphere.

The motor vehicle fleet was seen as a major source of hydrocarbon emissions and one which, if allowed to grow without some form of containment, would exacerbate photochemical smog in the major cities of Sydney and Melbourne. There was an awareness that tougher emission controls could increase vehicle fuel consumption and, although Australia had at that time indigenous crude oil resources to meet around 65-70% of domestic requirements, the price of imported oil was likely to continue to rise in real terms, with total supplies by no means guaranteed.

The first Australian Design Rules (ADR) for passenger vehicle emission standards - ADR 26 (limited carbon monoxide) and ADR 27 (limited hydrocarbons) - were introduced in the early 1970s. A more stringent standard, ADR 27A, equivalent to the US 1973 standard, covering exhaust emissions of nitrogen oxides, carbon monoxide and hydrocarbons, was implemented in 1976, but was found difficult and costly to administer and adversely affected vehicle fuel economy.

Two grades of leaded petrol were marketed during the 1970s; regular (89 Research Octane Number (RON)) and super (97 RON). The major proportion of sales was 97 RON with a lead content of 0.84 grams/litre.

Forecasts in the late 1970s indicated that ADR 27A would only contain photochemical pollution until the mid-1980s, after which increasing numbers of vehicles and deterioration in the emissions from older vehicles would exacerbate the problem.

In the late 1970s the Federal Government commissioned the Committee on Motor Vehicle Emissions (COMVE), comprising Federal and State officials, motor vehicle, oil industry and consumer representatives, to report on a strategy to combat atmospheric pollution.

COMVE began investigation of the competing issues of health, environment, energy conservation, cost to industry, lead times required, economic costs and benefits of different emission control strategies.

SELECTION OF APPROPRIATE EMISSION STANDARDS

Australia being a Federal system of six States and two Territories, the need for national uniformity in relation to a long term vehicle emissions strategy was recognised as an important consideration. The adoption of particular levels of emissions had national implications including the attainment of health, environment and energy objectives; lower costs for the motor vehicle industry and ultimately the consumer; economies in oil-refining operations; lower administrative costs; long term industry viability and effective transfer of technology.

There was concern that individual States (particularly New South Wales, because its capital, Sydney, was the most affected by photochemical smog) might adopt a go-it-alone strategy which would have adverse effects on Australian industry, consumers and government administration.

To determine the effect of tighter emission controls on the oil and motor vehicle industries, several strategies were considered. These ranged from:

- no change to the existing 1976 emission standards (hydrocarbons 1.9g/km, carbon monoxide 22g/km, nitrogen oxides 1.73 g/km) with the national average fuel consumption (NAFC) trending down to 8.5 litres/l00 km by 1986, to

- adopting US 1975 emission standards in 1986, with new cars using unleaded petrol (ULP), and

- meeting more stringent emission standards in 1991 or 1996, and national average fuel consumption to reach 8.2 litres/100 km in 1989.

OIL INDUSTRY

Annual estimates of refinery investment and operating costs for the period 1986-96 were derived from a linear programme refinery model used by the Federal Department of Primary Industries and Energy. Results showed that depending on the strategy option adopted, gross refinery investment would vary between $A 345 million and $A 376 million by 1996. The introduction of ULP would, and has had, little impact on refinery capacity. The model predicted that refinery operating costs would increase whatever the strategy adopted. For the ULP strategy 91.5 Research Octane Number (RON) - 82.5 Motor Octane Number (MON) was concluded to be the optimum petrol specification from a refinery operational viewpoint and vehicle industry attainability.

The oil industry expressed doubts about its ability to maintain an octane level of 91.5 RON, due to pressure to meet rising distillate demand. Octane capacity was not evenly distributed throughout the industry and some refiners required major plant addition to produce ULP without a large loss in total petrol output - 4 years lead time considered necessary to identify and justify optimal refinery process conditions, gain financial approval, design process plant and construct.

In spite of the oil industry's reticence about ULP production capacity, within 12 months of the introduction of the ULP emissions strategy in 1986, some refiners commenced marketing a higher (95 RON) premium unleaded petrol.

MOTOR VEHICLE INDUSTRY

It was recognised that the introduction of more stringent emission standards would impose costs on motor vehicle manufacturers and importers. In order to adopt US 1975 emission levels, vehicle manufacturers were expected to (and did) adopted catalyst technology to meet the requirements, as it became obvious that the "lean burn" approach would be generally unsuccessful, particularly for the larger engines.

COMVE concluded that the additional capital investment required by Australian vehicle manufacturers would be relatively small, with least cost for vehicles imported from Japan and assembled locally. The committee anticipated in 1980 (wrongly in the longer term) that forecast increases in crude oil and subsequently petrol prices, would assist the trend to smaller and lighter 4 cylinder vehicles.

The motor vehicle industry considered that the increased retail price of vehicles meeting the revised standards would slow sales, resulting in surplus capacity and lower employment levels. To some extent there was a reduction in new car sales in 1986-87 but stricter emission standards were only one of several factors contributing to the downturn. 1988 saw a recovery of the new car market to more buoyant levels, and more recently there has been another downturn. Emission standards are not considered as contributors to the more recent market fluctuations.

The change of petrol octane from 98 leaded to 91.5 unleaded was a considerable challenge, however, this was achieved by lowering engine compression ratios and installation of computerised engine management systems. All manufacturers have been able to meet the requirements. The use of fuel injection has now become common together with oxygen sensors and knock detectors.

Like the oil industry, the motor vehicle industry insisted on a lead time of around 5 years to redesign, develop, test and certify engines to operate on unleaded 91.5 RON fuel - the lead time was necessary to allow manufacturers to put maximum effort into minimising emissions and reducing fuel consumption, while maintaining acceptable performance in driveability, reliability and durability. The industry was assisted in that it had good access to overseas engine and catalyst technology.

THE UNLEADED PETROL STRATEGY

COMVE made its report to the Government in 1980 and in February 1981 the Australian Transport Advisory Council adopted a strategy which required all new vehicles manufactured after 1 January 1986 to be designed to operate on 91-93 octane ULP, to meet US 1975 emission standards, and for ULP to be available from 1 July 1985. Leaded petrol was to be phased out over a 20 year period. Changes to the emission standards required were incorporated in ADR 37, which is still the current standard.

All new vehicles marketed after 1 January 1986 were to operate on 91-93 RON ULP - this date was extended to 1 February 1986 to allow stocks of existing vehicles to be sold.

The use of catalyst technology to meet the emission standards was not prescribed, but overseas experience indicated that this would be the optimal route.

To discourage motorists from rendering emission control catalysts ineffective by misfuelling with leaded super petrol, petrol tank inlets on ULP cars are of smaller diameter than those on leaded petrol cars. In addition the Prices Surveillance Authority (a Commonwealth authority which sets wholesale prices for petrol) has been directed to set the wholesale price of ULP at or below the price of leaded petrol. The Commonwealth Government has no control over the retail prices of petrol; some State Governments have enacted legislation or issued guidelines to refiner/marketers requiring them to retail ULP at or below the price of leaded petrol. To date these methods of avoiding misfuelling have been very successful, with few reports of service station operators charging the incorrect price.

The distribution and general availability of ULP was assisted by removal of the regular (89 RON) leaded grade. Although many older cars were able to use regular leaded, few did, and the storage tanks at refineries and service stations were gradually flushed with ULP from July 1985. In isolated areas where only premium leaded had been sold ULP was made available in drums.

FUEL CONSUMPTION

Although Australia had good supplies of indigenous crude oil during the 1970s and was able to maintain a high degree of self sufficiency, the OPEC price rises of 1973 and 1979 led to concern that supplies of the heavier Middle East crudes used for lubricating oil basestocks, which Australia had to import, lacked security. Moreover, the knowledge that more stringent emission standards would lead to higher fuel consumption in motor vehicles stimulated the monitoring of the national fleet vehicle fuel consumption.

The National Average Fuel Consumption (NAFC) was calculated as a guide to changes over time of new vehicle fuel efficiency. NAFC is the sales weighted average of tested fuel consumptions of new passenger vehicles (except a few imported luxury vehicles) sold in Australia in a given year. The tests are conducted in accordance with Australian Standard 2877-1986. Combined with data on the rate of turnover of the vehicle fleet it provides a picture of longer term trends in vehicle fuel efficiency.

It is an imperfect indicator because changes in the pattern of market demand influence the NAFC. This has been particularly evident over the past two years, when the effect of some significant individual model improvements has been outweighed by a resurgence in demand for larger cars.

In 1978-79 the Federal Chamber of Automotive Industries, representing 98% of all vehicle manufacturers and importers, reached an agreement with the Government on progressive reductions in new passenger vehicle fuel consumption. At that time the NAFC was 11.5 litres/100 km. A target of 9 litres/100 km was set for 1983 and a preliminary target of 8.5 litres/100 km set for 1987-88. Neither target was met. The motor vehicle industry claimed that the tighter emission controls, together with the introduction of unleaded petrol, made achievement of the targets impossible. Furthermore, a resurgence of demand for six cylinder cars counterbalanced individual model improvements. In 1985-86, 27% of the new vehicle market was 6 cylinder cars; by 1988 this had risen to 33%.

Another important factor is that much of the design employed in the Australian motor vehicle industry is basically derivative, the introduction of major new development is slow because the market is relatively small and requires long model runs to recover development costs.

In 1987 a revised target of 8.7 litres/100 km was agreed for 1988 but this was also not achieved. The targets are set voluntarily - there are no penalties for non-achievement. The setting of voluntary targets for NAFC has now ceased and been replaced by a system of three year rolling forecasts which accompanies the industry's annual reporting.

The test procedure was changed in 1986 - the revised test figures are generally 2.75% lower in comparison with the old test procedure.

RESEARCH AND DEVELOPMENT PROGRAMME

Energy research, development and demonstration has played a major role in providing alternatives for a more balanced and diversified energy mix to ensure medium and long term energy supply security.

The main Commonwealth Government avenue for directing and coordinating energy RD&D is through the National Energy Research, Development and Demonstration (NERD&D) programme which was established in 1978. The programme has four key objectives:

- To support selectively Australian RD&D projects which have the potential to assist in meeting Government's energy policy objectives;
- Monitor and coordinate the Government's overall energy R&D effort;
- Disseminate results from projects supported under the NERD&D programme;
- Maximise technology transfer of NERD&D programme results to Australian industry.

An important area in which the Programme has made an active contribution to meeting energy policy objectives is in relation to alternative fuels development. Since its inception, around $5.7 million has been granted to passenger vehicle and engine technology projects. Recent research emphasis has been directed towards alternative fuels, compressed natural gas in particular, which will back out some petrol and distillate.

FUTURE DEVELOPMENTS

The motor vehicle industry, represented by the Federal Chamber of Automotive Industries (FCAI), has made application to ATAC to amend ADR 37 to allow certification of high performance, luxury vehicles on 95 RON octane fuel, since this grade has been adopted in Europe, Japan and the USA. This application will be considered at the March meeting of the Australian Transport Advisory Council.

Global concern about the possible effects of carbon dioxide emissions on climate change may lead to reduce emissions of carbon dioxide from motor vehicle exhausts. Any reduction is likely to be achieved by smaller, more efficient engines, improved transmission trains, better aerodynamics and smaller vehicle size and mass.

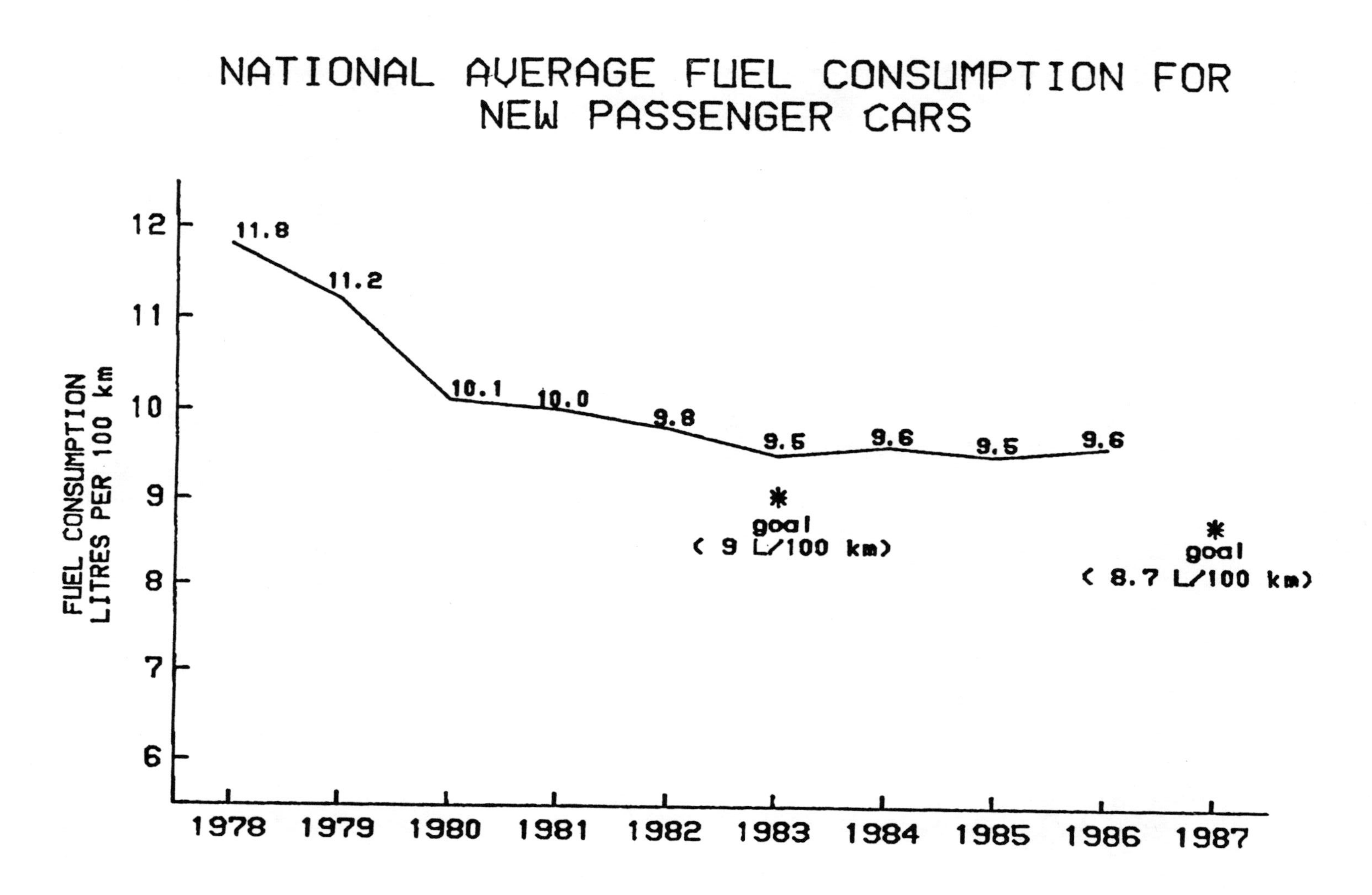
NATIONAL AVERAGE FUEL CONSUMPTION FOR
NEW PASSENGER CARS
FUEL CONSUMPTION
LITRES PER 100 km
12
11
10
9
8
7
6
11.8
11.2
10.1
10.0
9.8
9.5
9.6
9.5
9.6
goal
(9 L/100 km)
goal
(8.7 L/100 km)
1978
1979
1980
1981
1982
1983
1984
1985
1986
1987

APPENDIX I

FINAL PROGRAMME OF THE EXPERT PANEL MEETING

OECD / IEA INFORMAL EXPERT PANEL

ON

LOW CONSUMPTION / LOW EMISSION AUTOMOBILE

14TH-15TH FEBRUARY 1990

HOSTED BY ENEA - THE ITALIAN AGENCY FOR
NUCLEAR ENERGY AND ALTERNATIVE ENERGY RESOURCES

Viale Regina Margherita 125, Rome

Wednesday, 14th February

Chairman: Mr. G. Pinchera (Italy)

9.00 a.m.	**Welcome Address**	
	Prof. U. Colombo	President, ENEA
9.15 a.m.	**Opening Statements**	
	Mr. S. F. Garribba	Director, Office of Energy Technology, Research and Development, International Energy Agency (IEA)
	Mr. J. Short	Secretariat, European Conference of Ministers of Transport (ECMT)
	Mr. M. Roma	DG XVII, Commission of the European Communities (CEC)
	Mr. G. Dente	Economic Commission for Europe (ECE)
	Mr. C. Gerryn	Business and Industry Advisory Committee to the Organisation for Economic Co-operation and Development (BIAC)
	Mr. M. Hublin	Organisation internationale des constructeurs automobiles (OICA)
	Mr. G. Dorin	Environment Directorate, Organisation for Economic Co-operation and Development (OECD)
9.40 a.m.	**Tour de Table**	Brief Introduction of Experts

Wednesday, 14th February (Cont'd)

10.00 a.m. **Keynote Addresses**

- The automobile: environmental issues and energy security (challenges and opportunities for the OECD Member countries and others).

  Ms. D. L. Bleviss (International Institute for Energy Conservation, USA)

- Transport systems and the automobile in the urban context: problems of compatibility.

  Mr. R. Thörnblom (Transport Research Board, Sweden)

- National commitments and international policies for reduction of emissions.

  Mr. P. Stolpman (Environment Directorate, OECD)

10.40 a.m. COFFEE

11.10 a.m. **Overview of National Experiences**

Chairman: **Mr. G. Pinchera** (Italy)
Rapporteur: **Ms. D. L. Bleviss** (USA)

- Recent efforts and experience in reducing fuel consumption of automobiles while coping with environmental protection goals in selected countries.

- Brief statements from the following participants, followed by questions and discussion by experts.

  Mr. J. Brosthaus (Germany)
  Mr. L. Chinaglia (Italy)
  Mr. J. Delsey (France)
  Mr. C. Difiglio/Mr. J. Walsh (USA)
  Mr. M. Dunne/Mr. J. Bingham (UK)
  Mr. K. Egebäck (Sweden)
  Mr. K. Kontani (Japan)
  Mr. B. Rutten (The Netherlands)

1.30 p.m. LUNCH

Wednesday, 14th February (Cont'd)

3.00 p.m. **Overall Assessment of Major Constraints**

Chairman: **Mr. M. Walsh** (USA)

- Technological, industrial, marketing, regulatory and policy constraints in developing and promoting low consumption automobile on a large scale
- Introduction by Chairman followed by contributions from the participants to describe the current situation of automobile efficiency and emissions and to identify the major constraints to improving automobile fuel consumption and environmental impacts. Summary by Chairman of the key issues to be addressed in the remaining sessions.

4.15 p.m. COFFEE

4.45 p.m. **Technology Workshop I: Near-Term Options**

Chairman: **Mr. M. Walsh** (USA)
Rapporteur: **Mr. B. Rutten** (The Netherlands)

- Assessment of possible adaptation and modification of available technologies that would allow reduction in fuel consumption in the near-term (e.g. during the next five to seven years). Appraisal of cost-effective modifications likely to provide large gains (motor, transmission, electronics, aerodynamics, weight, etc.)
- Evaluation of technical potential of different options
- Assessment of environmental impact: atmospheric emission benefits in relation to energy efficiency gains.
- Factors likely to affect the development, production and market penetration of low consumption models
- Brief presentations on the above issues by panellists, based on their contributed papers, followed by discussion on these issues. Presentation of summary and conclusions by Chairman.

Panellists are:

Mr. J. R. Bang (Norway)
Mr. J. F. Bingham (UK)
Mr. L. Chinaglia (Italy)
Mr. J. Delsey (France)
Mr. K. G. Duleep (USA)
Mr. N. Gorßien (Germany)
Mr. M. Hublin (OICA)
Mr. T. Miyazaki (Japan)
Mr. L. O. Olsson (Sweden)
Mr. P. van Sloten (The Netherlands)

7.00 p.m. CLOSE OF SESSION

Thursday 15th February

9.00 a.m. **Economic / Industrial / Regulatory Aspects and Their Implications**

Chairman: **Ms. D. L. Blevis** (USA)
Rapporteur: **Mr. A. J. Bleijenberg** (The Netherlands)

- Review of significant aspects to be considered (commercial, economic, regulatory, energy, environmental) in relation to the main "actors" (industry, public and consumers, governments, international authorities, etc.)
- Environmental aspects: use of use of economic and regulatory incentives by governments (on vehicles, traffic, fuels, consumers, manufacturers) to hasten the penetration of low-emission vehicles.
- Implementation of appropriate regulations in OECD Member countries
- Bilateral or international mechanisms for accelerating trade and transfer of technology and experience to industry in non-OECD Member countries,
- Harmonisation of international environmental policies and standards
- Orientation of consumer choice towards more fuel-efficient vehicles
- Brief presentations on the above issues by panellists, based on their contributed papers, followed by discussion on these issues. Presentation of summary and conclusions by Chairman.

Panellists are:

Mr. G. Dente (ECE)
Mr. C. Difligio (USA)
Mr. C. Gerryn (BIAC)
Mr. N. Gorißen (Germany)
Mr. M. Hublin (OICA)
Mr. M. Kroon (Netherlands)
Ms. G. McInnes (IEA)
Mr. P. Reilly-Roe (Canada)
Mr. M. Roma (CEC, DG XVII)
Mr. J. Short (ECMT)
Mr. R. Thörnblom (Sweden)
Mr. M. Walsh (USA)

11.00 a.m. COFFEE

11.30 a.m. **Economic / Industrial / Regulatory Aspects and Their Implications** (Continued)

Chairman: **Mr. A. J. Bleijenberg** (The Netherlands)
Rapporteur: **Ms. D. L. Bleviss** (USA)

12.30 p.m. **Round Table: Identification of Key Issues and Relevant Short-Term Policies for the Attention of Governments**

- Round table discussion followed by Chairman's summary of conclusions on above issues.

1.30 p.m. LUNCH

Thursday 15th February (Cont'd)

2.45 p.m. **Technology Workshop II -- Medium/Long-Term Options**

Chairman: **Mr. G. Gerryn** (BIAC)
Rapporteur: **Mr. L. Chinaglia** (Italy)

- Potential for improvements in automobile efficiency and emissions by application of emerging technologies in the medium-term, e.g. advanced materials, information technology, alternative fuels

- New concepts under development or anticipated in the medium to long term: advanced engines and transmission, energy storage systems, substitute fuels, electric vehicles, changing duty cycle and utilisation patterns of automobiles and their impact on designs

- Limiting factors in efficiency and emissions improvement arising from present-generation automobile technology: e.g. anticipated maximum obtainable performance of reciprocating and rotary engines, internal combustion engines, conventional transmissions, available energy storage systems

- Managing the transition from present-generation to "next-wave" automobile technologies: identification of key technologies, assessment of present R&D activities, opportunities for enhanced effort and international collaboration.

- Brief presentations on the above issues by panellists, based on their contributed papers, followed by discussion on these issues. Presentation of summary and conclusions by Chairman.

Panellists are:

Mr. R. Alpaugh (USA)
Mr. Brosthaus (Germany)
Mr. J. Delsey (France)
Mr. K. G. Duleep (USA)
Mr. M. Hublin (OICA)
Mr. K. Kontani (Japan)
Mr. P. Reilly-Roe (Canada)
Mr. A. Rossi (CEC DG XII)
Mr. L. O. Olsson (Sweden)
Mr. C. Such (UK)
Mr. J. Van der Weide (Netherlands)

4.30 p.m. COFFEE

Thursday 15th February (Cont'd)

5.00 p.m. **Draft Plan of Future Actions**

Chairman: **Mr. J. Short** (ECMT)
Rapporteur: **Mr. D. Kearney** (IEA)

- Summary of issues and conclusions from previous sessions and technology workshops

- Brief review of programmes and timetables of national and international bodies (OECD/IEA, ECMT, IPCC, CEC, ECE) and prospects for further interaction

- Improved co-ordination of national and international efforts, including national and international assessments, information networks, proposals for collaboration on R&D and demonstration

- Opportunities for future work: establishment of special working groups on advanced energy technologies for road transport; structure and organisation of a seminar involving automobile manufacturers in order to discuss realistic approaches, practical time-scales and possible implementation of new solutions

- Brief presentations on the above issues by panellists, based on their contributed papers, followed by discussion on these issues. Presentation of summary and conclusions by Chairman.

  Panellists are:

  Ms. D. Bleviss (USA)
  Mr. G. Dente (ECE)
  Mr. S. F. Garribba (IEA)
  Mr. C. Gerryn (BIAC)
  Mr. M. Hublin (OICA)
  Mr. G. Pinchera (Italy)
  Mr. M. Roma (EEC)
  Mr. P. Stolpman (OECD)
  Mr. M. Walsh (USA)

6.30 p.m. **Closing Remarks**

- Mr. S. F. Garribba (Office of Energy Technology, Research and Development, IEA)

- Mr. P. Stolpman (Directorate for Environment, OECD)

7.00 p.m. **End of Expert Panel Meeting**

APPENDIX II

LIST OF REGISTERED PARTICIPANTS

AUSTRALIA

Mr. Peter Franklin
Counsellor
Agriculture & Minerals
Australian Embassy
Via Alessandria 215
00198 Rome, Italy

CANADA

Mr. Peter Reilly-Roe
Assistant Director
Transportation Energy Division
Energy Policy Sector
Energy, Mines & Resources
580 Booth Street
Ottawa, Ontario K1A OER

DENMARK

Mr. Nils Sogaard
Danish Technological Institute
Gregersensvej
Post Box 141
DK-2630 Taastrup

FINLAND

Prof. Antti Saarialho
Helsinki University of Technology
Otakaari 4
02150 Espoo

FRANCE

Mr. Jean Delsey
Institut National de Recherche
sur les Transports
109 av. Salvador Allende, Case 24
69675 Bron

Mr. Eyrat
Institut Français du Pétrole
1 ave. Bois Préau
BP 311
92506 Rueil Malmaison

Mr. M. Hublin
Organisation internationale
des constructeurs automobiles
4 rue de Berry
75008 Paris

FRANCE (Cont'd)

Mr. A. Morcheoine
Chef du Service Transport
Agence Française pour la
Maîtrise de l'Energie
Immeuble Béarn
27 rue Louis Vicat
75015 Paris

GERMANY

Mr. J. Brosthaus
TUV Rheinland
Institut fur Energie Technik
und Umweltschatz
Konstantin-Wille Strasse 1
500 Köln 91

Mr. Norbert Gorißen
Umweltbundesamt
Bismarckplatz 1
D-1000 Berlin 33

Mr. Dietmar Staschen
RDer
Federal Ministry of Economics
Villemombler Strasse 76
D-5300 Bonn

ITALY

Mr. Carlo Di Carlo
ENEA/FARE-UREN-TRA
CRE Casaccia, SP Anguillarese 301
00198 Rome

Mr. Leopoldo Chinaglia
Consiglio Scientifico
Istituto Motori, CNR
Strada Sappone 6
10133 Turin

Mr. Felice E. Corcione
Istituto Motori, CNR
Via Marconi 8
80125 Naples

Prof. Massimo Feola
Università di Roma II
Tor Vergata
00173 Rome

ITALY (Cont'd)

Mr. Giancarlo Pinchera
Consigliere Programmatico
Area Energia Ambiente e Salute
ENEA
Viale Regina Margherita 125
00198 Rome

Mr. Alberto Quaranta
Direttore, CNPM
Istituto Ricerche sulla Propulsione
e sull'Energetica del CNR
Via Francesco Baracca 69
20068 Peschiera Borromeo
Milan

JAPAN

Mr. Kazuo Kontani
Combustion Engineering Division
Mechanical Engineering Laboratory
Agency of Industrial Science
and Technology-MITI
2 Banci, Namiki 1-chome, Tsukuba-shi
Ibaragi-ken 305

Mr. Masatoshi Matsunami
Director-General
Office of Motor Vehicle
Pollution Control
Ministry of Transport
1-3, Kasumigaseki-2-chome
Chiyoda-ku
Tokyo 100

Mr. Takuro Miyazaki
Director
Office of Motor Vehicle Pollution
Control
Engineering and Planning Division
Ministry of Transport
1-3, Kasumigaseki-2-chome
Chiyoda-ku
Tokyo 100

NORWAY

Mr. Jon R. Bang
Senior Engineer
National Institute of Technology
PO Box 2608 St. Hanshaugen
N-0131 Oslo 1

NETHERLANDS

Mr. A. Bleijenberg
Center for Energy Conservation and
Environmental Technology
Oude Delft 180
2611 HH Delft

Mr. M. C. Kroon
Ministerie VROM
Directorate-General for
Environment
Ministry of Housing & Environment
PO BOX 450
2260 MB Leidschendam

Mr. B. J. C. M. Rutten
Center for Energy Conservation and
Environmental Technology
Oude Delft 180
2611 HH Delft

Mr. Peter van Sloten
TNO Road Vehicles
PO Box 237
2600 AE Delft

Mr. Jouke van der Weide
TNO Road Vehicles
PO Box 237
2600 AE Delft

SWEDEN

Mr. Karl-Erik Egebäck
Swedish Motor Vehicle
Inspection Company
Svensk Bilprovning
Kremlestigen 8
S-61163 Niköping

Mr. Lars Olou Olsson
Traffic Section
Environmental Protection Agency
S-17185 Solna

Mr. Ragnar Thörnblom
Swedish Traffic Research Board
Birger Jarls Torg 5
S-11128 Stockholm

UNITED KINGDOM

Mr. John F. Bingham
National Engineering Laboratory
East Kilbride
Glasgow G75 0QU
Scotland

Mr. Michael Bradley
Division of Air Quality
Dept. of Environment
Romney House B335
43 Marsham Street
London SW10 3 PY

Mr. J. M. Dunne
Head
Vehicle Emissions Group
Warren Spring Laboratory
Gunnels Wood Road
Stevenage
Hertfordshire SG12BX

Mr. M. Greening
Department of Transport
Office C19/03
Vehicle Standards and Engineering
2 Marsham Street
London SW1P 3EB

Mr. A. J. Hickman
Dept. of Transport
Vehicles & Environment Division
Transport and Road Research
Laboratory
Crowthorne
Berks RG11 GAV

Mr. Cecil Such
Manager, Client Services
Ricardo Consulting Engineers Plc
Bridge Works
Shoreham-by-Sea
West Sussex BN4 5FG

USA

Mr. Richard Alpaugh
Department of Energy
5G-046 FORS
1000 Independence Ave. SW
Washington DC 20585

USA (Cont'd)

Mr. Donald C. Bischoff
Director of Regulatory Analysis
National Highway Traffic Safety
Administration
400 7th Street SW
Washington DC 20590

Mrs. Deborah L. Bleviss
International Institute for Energy
Conservation
420 C Street NE
Washington DC 2002

Mr. Carmen Difiglio
Department of Energy
5G-046 FORS
1000 Independence Ave. SW
Washington DC 20585

Mr. K. G. Duleep
Energy & Environmental Analysis
Inc.
1655 North Ft. Meyer Drive,
Suite 600
Arlington, Virginia 22209

Mr. Frank von Hippel
Center for Energy and
Environmental Studies
Princeton University
Princeton, New Jersey 08544

Mr. Lee Schipper
International Energy Studies
Lawrence Berkeley Laboratory
Berkeley, California 94720

Mr. Jerry Stofflet
Motor Vehicle Manufacturers
Association
7430 Second Avenue
Suite 300
Detroit, Michigan 48202

Mr. Michael Walsh
2800 North Dinwiddie Street
Arlington, Virginia 22207

Commission of the European Communities (CEC)

Mr. Siegfried Heise
Commission of the European Communities
DG XI - Environment
200 rue de la Loi
B-1049 Brussels

Mr. Marcello Roma
Commission of the European Communities
DGXVII - Energy
200 rue de la Loi
B-1049 Brussels

Mr. Angelo Rossi
Commission of the European Communities
DG XII - Research & Technology
200 rue de la Loi
B-1049 Brussels

Economic Commission for Europe (ECE)

Mr. G. Dente
ECE
Palais des Nations
CH1211 Geneva

Business and Industry Advisory Committee to the OECD (BIAC)

Mr. Claude Gerryn
Environment and Safety
Liaison Manager
Bd. de la Wolowe 2
B-1150 Brussels

European Conference of Ministers of Transport (ECMT)

Mr. Jack Short
Transport Policy Division
ECMT
2, rue André-Pascal
75775 Paris

International Energy Agency (IEA)

Mr. Sergio Garribba
Director
Office of Energy Technology, Research and Development

Mr. Denis Kearney
Head of Division
Office of Energy Technology, Research and Development

Ms. Geneviève McInnes
Office of Long-Term Co-operation and Policy Analysis

Organisation for Economic Co-operation and Development (OECD), Directorate for Environment

Mr. Paul Stolpman
Head of Division
Pollution Control Division
Directorate for Environment

Mr. Gérard Dorin
Pollution Control Division
Directorate for Environment

Mr. Christian Averous
Economics Division
Directorate for Environment

Secretarial Staff

Ms. Nunzia Anselmo
ENEA
Viale Regina Margherita 125
00198 Rome, Italy

Ms. Margaret Jones
International Energy Agency
2 rue André-Pascal
75775 Paris, France

WHERE TO OBTAIN OECD PUBLICATIONS – OÙ OBTENIR LES PUBLICATIONS DE L'OCDE

Argentina – Argentine
Carlos Hirsch S.R.L.
Galería Güemes, Florida 165, 4° Piso
1333 Buenos Aires Tel. 30.7122, 331.1787 y 331.2391
Telegram: Hirsch–Baires
Telex: 21112 UAPE–AR. Ref. s/2901
Telefax:(1)331–1787

Australia – Australie
D.A. Book (Aust.) Pty. Ltd.
648 Whitehorse Road, P.O.B 163
Mitcham, Victoria 3132 Tel. (03)873.4411
Telex: AA37911 DA BOOK
Telefax: (03)873.5679

Austria – Autriche
OECD Publications and Information Centre
Schedestrasse 7
5300 Bonn 1 (Germany) Tel. (0228)21.60.45
Telefax: (0228)26.11.04

Gerold & Co.
Graben 31
Wien I Tel. (0222)533.50.14

Belgium – Belgique
Jean De Lannoy
Avenue du Roi 202
B–1060 Bruxelles Tel. (02)538.51.69/538.08.41
Telex: 63220 Telefax: (02) 538.08.41

Canada
Renouf Publishing Company Ltd.
1294 Algoma Road
Ottawa, ON K1B 3W8 Tel. (613)741.4333
Telex: 053–4783 Telefax: (613)741.5439
Stores:
61 Sparks Street
Ottawa, ON K1P 5R1 Tel. (613)238.8985
211 Yonge Street
Toronto, ON M5B 1M4 Tel. (416)363.3171

Federal Publications
165 University Avenue
Toronto, ON M5H 3B8 Tel. (416)581.1552
Telefax: (416)581.1743

Les Publications Fédérales
1185 rue de l'Université
Montréal, PQ H3B 3A7 Tel.(514)954–1633

Les Éditions La Liberté Inc.
3020 Chemin Sainte–Foy
Sainte–Foy, PQ G1X 3V6 Tel. (418)658.3763
Telefax: (418)658.3763

Denmark – Danemark
Munksgaard Export and Subscription Service
35, Nørre Søgade, P.O. Box 2148
DK–1016 København K Tel. (45 33)12.85.70
Telex: 19431 MUNKS DK Telefax: (45 33)12.93.87

Finland – Finlande
Akateeminen Kirjakauppa
Keskuskatu 1, P.O. Box 128
00100 Helsinki Tel. (358 0)12141
Telex: 125080 Telefax: (358 0)121.4441

France
OECD/OCDE
Mail Orders/Commandes par correspondance:
2 rue André–Pascal
75775 Paris Cedex 16 Tel. (1)45.24.82.00
Bookshop/Librairie:
33, rue Octave–Feuillet
75016 Paris Tel. (1)45.24.81.67
(1)45.24.81.81

Telex: 620 160 OCDE
Telefax: (33–1)45.24.85.00

Librairie de l'Université
12a, rue Nazareth
13090 Aix–en–Provence Tel. 42.26.18.08

Germany – Allemagne
OECD Publications and Information Centre
Schedestrasse 7
5300 Bonn 1 Tel. (0228)21.60.45
Telefax: (0228)26.11.04

Greece – Grèce
Librairie Kauffmann
28 rue du Stade
105 64 Athens Tel. 322.21.60
Telex: 218187 LIKA Gr

Hong Kong
Swindon Book Co. Ltd.
13 – 15 Lock Road
Kowloon, Hongkong Tel. 366 80 31
Telex: 50 441 SWIN HX
Telefax: 739 49 75

Iceland – Islande
Mál Mog Menning
Laugavegi 18, Pósthólf 392
121 Reykjavik Tel. 15199/24240

India – Inde
Oxford Book and Stationery Co.
Scindia House
New Delhi 110001 Tel. 331.5896/5308
Telex: 31 61990 AM IN
Telefax: (11)332.5993
17 Park Street
Calcutta 700016 Tel. 240832

Indonesia – Indonésie
Pdii–Lipi
P.O. Box 269/JKSMG/88
Jakarta 12790 Tel. 583467
Telex: 62 875

Ireland – Irlande
TDC Publishers – Library Suppliers
12 North Frederick Street
Dublin 1 Tel. 744835/749677
Telex: 33530 TDCP EI Telefax : 748416

Italy – Italie
Libreria Commissionaria Sansoni
Via Benedetto Fortini, 120/10
Casella Post. 552
50125 Firenze Tel. (055)645415

Telex: 570466 Telefax: (39.55)641257
Via Bartolini 29
20155 Milano Tel. 365083
La diffusione delle pubblicazioni OCSE viene assicurata dalle principali librerie ed anche da:
Editrice e Libreria Herder
Piazza Montecitorio 120
00186 Roma Tel. 679.4628
Telex: NATEL I 621427

Libreria Hoepli
Via Hoepli 5
20121 Milano Tel. 865446
Telex: 31.33.95 Telefax: (39.2)805.2886

Libreria Scientifica
Dott. Lucio de Biasio "Aeiou"
Via Meravigli 16
20123 Milano Tel. 807679
Telefax: 800175

Japan– Japon
OECD Publications and Information Centre
Landic Akasaka Building
2–3–4 Akasaka, Minato–ku
Tokyo 107 Tel. (81.3)3586.2016
Telefax: (81.3)3584.7929

Korea – Corée
Kyobo Book Centre Co. Ltd.
P.O. Box 1658, Kwang Hwa Moon
Seoul Tel. (REP)730.78.91
Telefax: 735.0030

Malaysia/Singapore – Malaisie/Singapour
Co–operative Bookshop Ltd.
University of Malaya
P.O. Box 1127, Jalan Pantai Baru
59700 Kuala Lumpur
Malaysia Tel. 756.5000/756.5425
Telefax: 757.3661

Information Publications Pte. Ltd.
Pei–Fu Industrial Building
24 New Industrial Road No. 02–06
Singapore 1953 Tel. 283.1786/283.1798
Telefax: 284.8875

Netherlands – Pays–Bas
SDU Uitgeverij
Christoffel Plantijnstraat 2
Postbus 20014
2500 EA's–Gravenhage Tel. (070 3)78.99.11
Voor bestellingen: Tel. (070 3)78.98.80
Telex: 32486 stdru Telefax: (070 3)47.63.51

New Zealand – Nouvelle–Zélande
Government Printing Office
Customer Services
33 The Esplanade – P.O. Box 38–900
Petone, Wellington
Tel. (04) 685–555 Telefax: (04)685–333

Norway – Norvège
Narvesen Info Center – NIC
Bertrand Narvesens vei 2
P.O. Box 6125 Etterstad
0602 Oslo 6 Tel. (02)57.33.00
Telex: 79668 NIC N Telefax: (02)68.19.01

Pakistan
Mirza Book Agency
65 Shahrah Quaid–E–Azam
Lahore 3 Tel. 66839
Telex: 44886 UBL PK. Attn: MIRZA BK

Portugal
Livraria Portugal
Rua do Carmo 70–74
Apart. 2681
1117 Lisboa Codex Tel. 347.49.82/3/4/5
Telefax: 37 02 64

Singapore/Malaysia – Singapour/Malaisie
See "Malaysia/Singapore – "Voir "Malaisie/Singapour"

Spain – Espagne
Mundi–Prensa Libros S.A.
Castelló 37, Apartado 1223
Madrid 28001 Tel. (91) 431.33.99
Telex: 49370 MPLI Telefax: 575 39 98
Libreria Internacional AEDOS
Consejo de Ciento 391
08009 –Barcelona Tel. (93) 301–86–15
Telefax: (93) 317–01–41

Sweden – Suède
Fritzes Fackboksföretaget
Box 16356, S 103 27 STH
Regeringsgatan 12
DS Stockholm Tel. (08)23.89.00
Telex: 12387 Telefax: (08)20.50.21

Subscription Agency/Abonnements:
Wennergren–Williams AB
Nordenflychtsvagen 74
Box 30004
104 25 Stockholm Tel. (08)13.67.00
Telex: 19937 Telefax: (08)618.62.36

Switzerland – Suisse
OECD Publications and Information Centre
Schedestrasse 7
5300 Bonn 1 (Germany) Tel. (0228)21.60.45
Telefax: (0228)26.11.04

Librairie Payot
6 rue Grenus
1211 Genève 11 Tel. (022)731.89.50
Telex: 28356
Subscription Agency – Service des Abonnements
4 place Pépinet – BP 3312
1002 Lausanne Tel. (021)341.33.31
Telefax: (021)341.33.45

Maditec S.A.
Ch. des Palettes 4
1020 Renens/Lausanne Tel. (021)635.08.65
Telefax: (021)635.07.80
United Nations Bookshop/Librairie des Nations–Unies
Palais des Nations
1211 Genève 10 Tel. (022)734.60.11 (ext. 48.72)
Telex: 289696 (Attn: Sales)
Telefax: (022)733.98.79

Taiwan – Formose
Good Faith Worldwide Int'l. Co. Ltd.
9th Floor, No. 118, Sec. 2
Chung Hsiao E. Road
Taipei Tel. 391.7396/391.7397
Telefax: (02) 394.9176

Thailand – Thaïlande
Suksit Siam Co. Ltd.
1715 Rama IV Road, Samyan
Bangkok 5 Tel. 251.1630

Turkey – Turquie
Kültur Yayinlari Is–Türk Ltd. Sti.
Atatürk Bulvari No. 191/Kat. 21
Kavaklidere/Ankara Tel. 25.07.60
Dolmabahce Cad. No. 29
Besiktas/Istanbul Tel. 160.71.88
Telex: 43482B

United Kingdom – Royaume–Uni
HMSO
Gen. enquiries Tel. (071) 873 0011
Postal orders only:
P.O. Box 276, London SW8 5DT
Personal Callers HMSO Bookshop
49 High Holborn, London WC1V 6HB
Telex: 297138 Telefax: 071 873 8463
Branches at: Belfast, Birmingham, Bristol, Edinburgh, Manchester

United States – États–Unis
OECD Publications and Information Centre
2001 L Street N.W., Suite 700
Washington, D.C. 20036–4095 Tel. (202)785.6323
Telefax: (202)785.0350

Venezuela
Libreria del Este
Avda F. Miranda 52, Aptdo. 60337
Edificio Galipán
Caracas 106 Tel. 951.1705/951.2307/951.1297
Telegram: Libreste Caracas

Yugoslavia – Yougoslavie
Jugoslovenska Knjiga
Knez Mihajlova 2, P.O. Box 36
Beograd Tel. (011)621.992
Telex: 12466 jk bgd Telefax: (011)625.970

Orders and inquiries from countries where Distributors have not yet been appointed should be sent to: OECD Publications Service, 2 rue André–Pascal, 75775 Paris Cedex 16, France.
Les commandes provenant de pays où l'OCDE n'a pas encore désigné de distributeur devraient être adressées à : OCDE, Service des Publications, 2, rue André–Pascal, 75775 Paris Cedex 16, France.

12/90

OECD PUBLICATIONS, 2, rue André-Pascal, 75775 PARIS CEDEX 16 - N° 45486 1991
PRINTED IN FRANCE
(61 91 03 1) ISBN 92-64-13465-4